FEDERAL AVIATION ADMINISTRATION AIRSPACE REDESIGN AND CONGESTION MANAGEMENT

FEDERAL AVIATION ADMINISTRATION AIRSPACE REDESIGN AND CONGESTION MANAGEMENT

SIDNEY R. ROTHMAN
EDITOR

Nova Science Publishers, Inc.
New York

For permission to use material from this book please contact us:
Telephone 631-231-7269; Fax 631-231-8175
Web Site: http://www.novapublishers.com

LIBRARY OF CONGRESS CATALOGING-IN-PUBLICATION DATA

FAA airspace redesign and congestion management / editor, Sidney R. Rothman.
p. cm.
Includes index.
ISBN 978-1-60692-710-6 (hardcover)
1. Air traffic control--Technological innovations--United States. 2. Air traffic control--Government policy--United States. 3. Air traffic capacity--United States. 4. Airport capacity--United States. 5. United States. Federal Aviation Administration. I. Rothman, Sidney R.
TL725.3.T7F33 2009
387.7'404260973--dc22

2009028454

Published by Nova Science Publishers, Inc. ✝ New York

CONTENTS

PREFACE

Chapter 1 - FAA has not developed a detailed implementation plan for the New York/New Jersey/Philadelphia Airspace Redesign with a schedule, and therefore GAO was unable to determine whether FAA would meet its projected timetable. In addition, the final project configuration and costs are unknown since FAA has not determined the type of equipment and software that will be needed and FAA is currently reviewing whether to house operations for the redesigned airspace in existing FAA facilities, a new facility, or a consolidated facility. Given that the redesign represents a complex and comprehensive change to the region's airspace, GAO believes it is important to conduct evaluations of the redesign after each implementation step to ensure proper implementation. A potential strategy that could be used by FAA is an adaptive management strategy, which is a process that promotes flexible decision making as outcomes from management actions become understood.

Chapter 2 - Aviation congestion and delay is an issue of long standing. Statistically, there was a significant increase in congestion and delay throughout the national aviation system in 2007. The situation has been especially noticeable at certain key airports, namely the New York region in general and John F. Kennedy International Airport (JFK) in particular. The Department of Transportation (DOT) and its operating agency, the Federal Aviation Administration (FAA), have promised to take actions aimed at reducing congestion and delay both in the short and long terms. In the short term, DOT is proposing to address the JFK situation through administrative and economic measures that would likely restrict or otherwise provide for the allocation of flights into the airport during specific periods of time.

Chapter 3 – This is a statement of James C. May, President and Chief Executive Officer of Air Transport Association of America, Inc. before the Subcommittee on Aviation of the Committee on Transportation and Infrastructure: Congestion Management in the New York Airspace.

Chapter 4 – This is a testimony of Douglas E. Lavin, Regional Vice President (North America), International Air Transport Association, Committee on Transportation and Infrastructure's Subcommittee on Aviation, United States House of Representatives.

Chapter 5 – This is a testimony of William R. Decota, Aviation Director, The Port Authority of New York and New Jersey: United States House of Representatives, Transportation and Infrastructure Committee, Subcommittee on Aviation.

Chapter 6 – This is a testimony of Edward P. Faberman, Executive Director, Air Carrier Association of America before the United States House of Representatives, Committee on Transportation and Infrastructure, Subcommittee on Aviation Regarding "Congestion Management in the New York Airspace".

Chapter 7 – This is a testimony of Congressman Christopher Shays "Congestion Management in the New York Airspace", Subcommittee on Aviation, House Committee on Transportation and Infrastructure.

Chapter 8 – These are remarks as prepared of United States Senator Charles E. Schumer, House Subcommittee on Aviation: Congestion Management in the New York Airspace.

Chapter 9 – This is a statement of the Honorable D.J. Gribbin General Counsel, United States Department of Transportation before the United States House of Representatives Committee on Transportation and Infrastructure Subcommittee on Aviation Concerning Aviation Congestion Management.

In: FAA Airspace Redesign and Congestion Management ISBN: 978-1-60692-710-6
Editor: Sidney R. Rothman

Chapter 1

FEDERAL AVIATION ADMINISTRATION AIRSPACE REDESIGN: AN ANALYSIS OF THE NEW YORK/NEW JERSEY/PHILADELPHIA PROJECT

Government Accountability Office

Why GAO Did This Study

In September 2007, after 9 years of evaluation and a cost of over $53 million, the Federal Aviation Administration (FAA) announced it would begin implementing a new airspace structure for the New York/New Jersey/Philadelphia metropolitan area. According to FAA, this redesign of routes leading to and from commercial airports will fully integrate the airspace in the region, produce $300 million annual savings, and reduce delay by 20 percent once fully implemented. Critics disagree and cite potential increases in aircraft noise and other adverse environmental impacts. GAO was asked to examine: (1) the extent to which FAA followed legal requirements for its environmental review, (2) the extent to which FAA's methodology in assessing operational and noise impacts was reasonable, and (3) the likelihood FAA will meet its projected time frames and costs of implementation. GAO's legal analysis covered applicable federal laws, regulations, court decisions, and FAA orders. GAO's analysis of FAA's methodology was based on criteria established through review of federal policy, FAA's guidance, prior GAO reports, and standards from the aviation and analytical community. With the assistance of the National Academy of Sciences, GAO identified experts in the fields of environmental policies and procedures, airspace operations, and aircraft noise measurement and obtained their views on relevant aspects of FAA's methodology.

GAO evaluated FAA's compliance with the National Environmental Policy Act (NEPA) and environmental justice directives in conducting the New York/New Jersey/Philadelphia Airspace Redesign project. In assessing compliance, GAO used established court precedent applying these requirements, as well as the standard of review for agency actions established by the Administrative Procedure Act (APA), which is deferential to agency decision making. Courts interpret the APA standard—whether an agency's actions were "arbitrary, capricious, an abuse of discretion, or otherwise not in accordance with law"—as mandating that an agency act reasonably in carrying out NEPA's requirements and that the agency's ultimate decisions be reasonable and not arbitrary and capricious. GAO reviewed FAA's compliance with respect to five key issues: the statement of the project's purpose and need, the evaluation of alternatives, consideration of the project's environmental effects, public participation, and environmental justice matters. GAO selected these issues based on public concerns raised during and after the NEPA process, congressional interest, the views of experts we interviewed, and GAO's evaluation of the range of concerns presented.

Applying these legal requirements and the APA's reasonableness standard, GAO concluded that FAA complied with applicable NEPA requirements and related environmental justice directives. First, the statement of the project's purpose and need—which defines the objective of the project and which, in this case, was to increase the efficiency and reliability of the airspace while enhancing safety and reducing delays—was reasonable. The statement was reasonable in scope, as it was not defined too narrowly or too broadly, and it reasonably excluded noise reduction. Second, FAA developed a reasonable range of alternatives to the redesign and appropriately evaluated these alternatives. As required, FAA included a no-action alternative to serve as a baseline, as well as alternatives that would achieve the project's purpose and need. FAA also discussed options eliminated from detailed analysis, and explored and objectively evaluated the remaining alternatives. Third, FAA acted reasonably in not analyzing the indirect environmental effects of potential growth resulting from the redesign. Because FAA found the redesign in itself would not increase traffic demand and flight operations, it did not consider the potential environmental impacts of these system improvements. In the aviation context, courts have uniformly upheld similar decisions by FAA where, as in this case, the purpose of the project was not to induce growth and the project did not include capacity-enhancing construction, such as the addition of a runway. Fourth, FAA reasonably involved the public throughout the environmental review process. It took actions required to ensure public outreach including conducting an early and open process, providing notice of and holding public meetings, and soliciting and responding to public comments. Fifth, FAA satisfied environmental justice directives in Executive Order 12898 and related guidance and Orders. FAA prepared an analysis that identified minority and low-income populations significantly impacted by the proposed redesign, and determined whether the impacton these populations was disproportionate. FAA also involved these individuals throughout the environmental review process. In addition, FAA mitigated these significant impacts by altering arrival procedures and departure headings, raising arrival altitudes, and other related measures.

FAA's methodology to assess operational and noise impacts was reasonable, based on FAA's guidance for conducting airspace redesigns, standards from the aviation and analytical community, and the opinion of independent aviation operational and noise experts. FAA's guidance suggests activities for conducting a redesign's operational analysis and establishes specific guidelines for conducting a noise analysis. FAA generally adhered to this guidance in conducting the redesign, in that FAA generally followed its process for conducting operational analyses and used the noise modeling tool and metric specified in its guidance. In addition, according to experts, FAA used experienced contractors, the best available modeling tools, and appropriate data. For example, according to FAA and experts we interviewed, the data sources, such as FAA radar flight track data, U.S. Geological Survey terrain data, and U.S. Census Bureau data, are industry standard and generally recognized as providing reliable information. However, GAO and experts also identified some ways in which the methodology could be improved for future redesign projects. For example, when evaluating the alternatives, FAA did not analyze various economic impacts, such as implementation costs. GAO identified two types of analyses—an uncertainty analysis and a benefit-cost analysis—that could have benefited decision makers and the public in future redesign efforts. An uncertainty analysis would provide more information about the level of uncertainty associated with conducting the operational analysis, while a benefit-cost analysis would provide more information about the impacts of various alternatives.

FAA has not developed a detailed implementation plan for the New York/New Jersey/Philadelphia Airspace Redesign with a schedule, and therefore GAO was unable to determine whether FAA would meet its projected timetable. In addition, the final project configuration and costs are unknown since FAA has not determined the type of equipment and software that will be needed and FAA is currently reviewing whether to house operations for the redesigned airspace in existing FAA facilities, a new facility, or a consolidated facility. Given that the redesign represents a complex and comprehensive change to the region's airspace, GAO believes it is important to conduct evaluations of the redesign after each implementation step to ensure proper implementation. A potential strategy that could be used by FAA is an adaptive management strategy, which is a process that promotes flexible decision making as outcomes from management actions become understood.

What GAO Recommends

GAO recommends that FAA develop and follow a detailed implementation plan for the New York/New Jersey/Philadelphia Airspace Redesign that includes a time and cost schedule and follow a post implementation evaluation plan that includes an adaptive management strategy. In addition, GAO recommends that for future airspace redesign projects, FAA conduct an uncertainty analysis of key assumptions and inputs to provide information on the level of confidence of the estimated impacts and conduct a benefit-cost analysis for the purpose of evaluating redesign alternatives.

A draft was provided to DOT and FAA. While DOT said it does not agree with everything in the draft, due to pending litigation, DOT declined to specify its areas of agreement or disagreement. DOT provided technical comments that we incorporated as appropriate.

Abbreviations

APA Administrative Procedure Act
ATC Air Traffic Control
ATO Air Traffic Organization
CAASD Center for Advanced Aviation System Development
CEQ Council on Environmental Quality
DNL Day-Night Average Sound Level
DOT Department of Transportation
EIS environmental impact statement
FAA Federal Aviation Administration
FFRDC Federally Funded Research and Development Center
FHWA Federal Highway Administration
FICON Federal Interagency Committee on Noise
Handbook FAA Airspace Management Handbook
ICC Integrated Control Complex
JFK John F. Kennedy International Airport
LaGuardia LaGuardia Airport
NATCA National Air Traffic Controllers Association
NEPA National Environmental Policy Act
Newark Newark Liberty International Airport
NextGen Next Generation Air Transportation System
NIRS Noise Integrated Routing System
NJCAAN New Jersey Coalition Against Aircraft Noise
Philadelphia Philadelphia International Airport
RNAV Area Navigation (technology)
SEL Sound Exposure Level
TAAM Total Airspace and Airport Modeler
Teterboro Teterboro Airport
TRACON Terminal Radar Approach Control

What GAO Found

GAO evaluated FAA's compliance with the National Environmental Policy Act (NEPA) and environmental justice directives in conducting the New York/New Jersey/Philadelphia Airspace Redesign project. In assessing compliance, GAO used established court precedent applying these requirements, as well as the standard of review for agency actions established by the Administrative Procedure Act (APA), which is deferential to agency decision making. Courts interpret the APA standard—whether an agency's actions were "arbitrary, capricious, an abuse of discretion, or otherwise not in accordance with law"—as mandating that an agency act reasonably in carrying out NEPA's requirements and that the agency's ultimate decisions be reasonable and not arbitrary and capricious. GAO reviewed FAA's compliance with respect to five key issues: the statement of the project's purpose and need, the evaluation of alternatives, consideration of the project's environmental effects, public participation, and environmental justice matters. GAO selected these issues based on public concerns raised during and after the NEPA process, congressional interest, the views of experts we interviewed, and GAO's evaluation of the range of concerns presented.

Applying these legal requirements and the APA's reasonableness standard, GAO concluded that FAA complied with applicable NEPA requirements and related environmental justice directives. First, the statement of the project's purpose and need—which defines the objective of the project and which, in this case, was to increase the efficiency and reliability of the airspace while enhancing safety and reducing delays—was reasonable. The statement was reasonable in scope, as it was not defined too narrowly or too broadly, and it reasonably excluded noise reduction. Second, FAA developed a reasonable range of alternatives to the redesign and appropriately evaluated these alternatives. As required, FAA included a no-action alternative to serve as a baseline, as well as alternatives that would achieve the project's purpose and need. FAA also discussed options eliminated from detailed analysis, and explored and objectively evaluated the remaining alternatives. Third, FAA acted reasonably in not analyzing the indirect environmental effects of potential growth resulting from the redesign. Because FAA found the redesign in itself would not increase traffic demand and flight operations, it did not consider the potential environmental impacts of these system improvements. In the aviation context, courts have uniformly upheld similar decisions by FAA where, as in this case, the purpose of the project was not to induce growth and the project did not include capacity-enhancing construction, such as the addition of a runway. Fourth, FAA reasonably involved the public throughout the environmental review process. It took actions required to ensure public outreach including conducting an early and open process, providing notice of and holding public meetings, and soliciting and responding to public comments. Fifth, FAA satisfied environmental justice directives in Executive Order 12898 and related guidance and Orders. FAA prepared an analysis that identified minority and low-income populations significantly impacted by the proposed redesign, and determined whether the impacton these populations was disproportionate. FAA also involved these individuals throughout the environmental review process. In addition, FAA mitigated these significant impacts by altering arrival procedures and departure headings, raising arrival altitudes, and other related measures.

FAA's methodology to assess operational and noise impacts was reasonable, based on FAA's guidance for conducting airspace redesigns, standards from the aviation and analytical community, and the opinion of independent aviation operational and noise experts. FAA's guidance suggests activities for conducting a redesign's operational analysis and establishes specific guidelines for conducting a noise analysis. FAA generally adhered to this guidance in conducting the redesign, in that FAA generally followed its process for conducting operational analyses and used the noise modeling tool and metric specified in its guidance. In addition, according to experts, FAA used experienced contractors, the best available modeling tools, and appropriate data. For example, according to FAA and experts we interviewed, the data sources, such as FAA radar flight track data, U.S. Geological Survey terrain data, and U.S. Census Bureau data, are industry standard and generally recognized as providing reliable information. However, GAO and experts also identified some ways in which the methodology could be improved for future redesign projects. For example, when evaluating the alternatives, FAA did not analyze various economic impacts, such as implementation costs. GAO identified two types of analyses—an uncertainty analysis and a benefit-cost analysis—

that could have benefited decision makers and the public in future redesign efforts. An uncertainty analysis would provide more information about the level of uncertainty associated with conducting the operational analysis, while a benefit-cost analysis would provide more information about the impacts of various alternatives.

FAA has not developed a detailed implementation plan for the New York/New Jersey/Philadelphia Airspace Redesign with a schedule, and therefore GAO was unable to determine whether FAA would meet its projected timetable. In addition, the final project configuration and costs are unknown since FAA has not determined the type of equipment and software that will be needed and FAA is currently reviewing whether to house operations for the redesigned airspace in existing FAA facilities, a new facility, or a consolidated facility. Given that the redesign represents a complex and comprehensive change to the region's airspace, GAO believes it is important to conduct evaluations of the redesign after each implementation step to ensure proper implementation. A potential strategy that could be used by FAA is an adaptive management strategy, which is a process that promotes flexible decision making as outcomes from management actions become understood.

July 31, 2008

The Honorable James Oberstar
Chairman
Committee on Transportation and Infrastructure
House of Representatives

The Honorable Jerry Costello
Chairman
Subcommittee on Aviation
Committee on Transportation and Infrastructure
House of Representatives

The Honorable Robert Andrews
The Honorable Joseph Sestak
House of Representatives

In September 2007, after 9 years of evaluation and a cost of over $53 million, the Federal Aviation Administration (FAA) announced that it would implement a new airspace structure for the five major airports[1] and several regional airports serving the New York/New Jersey/Philadelphia metropolitan area.[2] According to FAA, the expected growth in the region's air travel contributed to the need for the airspace redesign, the goal of which is to increase the efficiency and reliability of the region's airspace, while maintaining safety. The efficient operation of the airspace is important, since the region's airports are critical to the nation's aviation system. For example, because the New York metropolitan area airports provide service to one of the most populated urban areas in the United States, as well as to substantial commercial air traffic,[3] any delays in this of region tend to ripple throughout the National Airspace System. The region's airspace, therefore, has become a critical chokepoint, and any resulting problems have substantial national impacts. FAA believes that the redesigned airspace will have several important benefits, including reducing airport delays, increasing the balance of air traffic controller workloads and operational flexibility, and maintaining the throughput of airports. FAA estimates that when compared to the current airspace structure, the redesigned airspace will produce $300 million annual savings in direct costs and will reduce delay by 20 percent once fully implemented.[4] However, many in the region's local communities disagree While this litigation is pending, FAA has begun implementation that the redesign will yield these benefits and worry that the redesign will substantially increase aircraft noise and produce other environmental impacts, both in communities already affected and in areas where aircraft traditionally have not flown. A number of critics maintain that FAA's decision to implement the new airspace structure is based on a flawed process and a predetermined outcome. For example, some critics question the methodology FAA used to develop the different alternatives evaluated in the airspace redesign and maintain that FAA did not fully assess nonairspace alternatives, such as requiring airlines to pay additional fees for airport use during peak demand times. Based on these and other objections, state and local governments and citizen groups have filed 13 separate lawsuits challenging the redesign, based primarily on FAA's alleged failure to comply with the National Environmental Policy Act (NEPA).[5] While this litigation is pending, FAA has begun implementation of the redesign, with initial steps starting on December 19, 2007.[6] Full implementation is estimated to be completed by 2012.

You asked us to review FAA's airspace redesign project for the New York/New Jersey/Philadelphia region. Accordingly, this chapter focuses on the following questions:

1. To what extent did FAA follow key legal procedures and requirements in conducting its environmental review?
2. To what extent was the methodology used by FAA to assess the operational and noise impacts reasonable?
3. What is the likelihood that FAA will meet its project time frames and costs of implementing its airspace redesign project?

To address our three research questions, we reviewed documents associated with the New York/New Jersey/Philadelphia Airspace Redesign, including the draft and final environmental impact statements (EIS) and selected appendices, underlying technical reports and analyses, and the administrative record filed by FAA with the court in the pending litigation. We also obtained and analyzed information from a variety of other sources, including previous FAA airspace redesigns similar to this project and other aviation evaluations and studies.[7] We interviewed officials from the Department of Transportation (DOT) and from FAA's headquarters and eastern regional office, including representatives of FAA's airspace redesign team. We also contacted other stakeholders, including FAA's principal contractors[8] for the operational and noise analyses, the Port Authority of New York and New Jersey, City of Philadelphia Department of Aviation, other participants in the EIS process, trade organizations for the aviation industry, the national air traffic controller's union, and community groups in the region.

To assess the extent to which FAA followed key legal procedures and requirements in conducting its environmental review, we reviewed applicable federal laws, regulations, executive directives, and court decisions, as well as FAA and DOT Orders and FAA and Council on Environmental Quality (CEQ) guidance. We then conducted a legal analysis to determine FAA's compliance with respect to five key issues: the statement of the project's purpose and need, evaluation of alternatives, consideration of the project's environmental effects, public involvement, and environmental justice matters. We selected these five issues based on public concerns raised during and after the EIS process,[9] congressional interest, the views of experts we interviewed, and GAO's evaluation of the range of concerns presented. To examine the extent to which the methodology used by FAA to assess the operational and noise impacts was reasonable, we compared FAA's methodology to criteria we established through review of federal policy, FAA guidance, prior GAO reports, and standards from the aviation and analytical community. In addition, with the assistance of the National Academy of Sciences, we identified experts in the fields of EIS policies and procedures, airspace operations and system modeling, and aircraft noise measurement and mitigation. These experts reviewed selected portions of the EIS related to FAA's operational and noise analysis. We then interviewed these experts within their area of expertise to obtain their views on the extent to which FAA followed applicable procedures and requirements, and on the methodology used by FAA to assess the operational and noise impacts. See appendix I for a list of the experts we interviewed.

As agreed with you, we did not prepare a new EIS or develop and analyze new alternatives to the airspace redesign. We conducted this performance audit from July 2007 to July 2008 in accordance with generally accepted government auditing standards. Those standards require that we plan and perform the audit to obtain sufficient, appropriate evidence to provide a reasonable basis for our findings and conclusions based on our audit objectives. We believe that the evidence obtained provides a reasonable basis for our findings and conclusions based on our audit objectives.

RESULTS IN BRIEF

Our evaluation of FAA's compliance with NEPA in conducting the New York/New Jersey/Philadelphia Airspace Redesign was based on established court precedent applying NEPA and its implementing regulations, and the standard of review for agency actions established by the Administrative Procedure Act (APA), which is deferential to agency decision making. Courts interpret this APA standard—whether an agency's actions were "arbitrary, capricious, an abuse of discretion, or otherwise not in accordance with law"—as requiring that an agency act reasonably in carrying out NEPA's requirements and that the agency's ultimate decisions be reasonable and not arbitrary and capricious. We reviewed FAA's compliance with respect to five key issues: the statement of the project's purpose and need, evaluation of alternatives, consideration of the project's environmental effects, public involvement, and environmental justice matters. We selected these five issues based on public concerns raised during and after the EIS process, congressional interest, the views of experts we interviewed, and GAO's evaluation of the range of concerns presented.

Applying these NEPA requirements and the APA's reasonableness standard to these five key issues, we concluded that FAA complied with applicable NEPA and related requirements and environmental justice directives. First, the statement of the project's purpose and need—which defines the objectives of the project and which, in this case, was to increase the efficiency and reliability of the airspace while maintaining safety and reducing delays—was reasonable. The statement was reasonable in scope because it was not defined too narrowly or too broadly, and it reasonably excluded noise reduction. Second, FAA developed a reasonable range of alternatives to the redesign and appropriately evaluated these alternatives. As required, FAA included a no action alternative, which served as the baseline, as well as alternatives that would achieve the project's purpose and need. FAA also discussed options it eliminated from detailed analysis, and explored and objectively evaluated the remaining alternatives. Third, FAA acted reasonably in not analyzing the indirect environmental effects of potential growth resulting from the redesign. Because FAA found that the redesign in itself would not increase traffic demand and flight operations, it did not consider the potential environmental impacts of these system improvements. In the aviation context, the courts have uniformly upheld similar decisions by FAA where, as in this case, the purpose of the project was not to induce growth and the project did not include capacity-enhancing construction, such as the addition of a runway. Fourth, FAA reasonably involved the public throughout the environmental review process. It took the actions required to ensure appropriate public outreach, including conducting an early and open process, providing notice of and holding public meetings, and soliciting and responding to public comments. Fifth, FAA satisfied environmental justice directives in Executive Order 12898 and implementing CEQ guidance and DOT Orders. FAA prepared an analysis that identified minority and low-income populations significantly impacted by the proposed redesign and determined whether the impact on these populations was disproportionate. FAA also involved these individuals throughout the environmental review process. In addition, FAA mitigated these significant impacts by altering arrival procedures and departure headings, raising arrival altitudes, and other related measures.

FAA's methodology to assess the operational and noise impacts was reasonable, based on FAA's guidance for conducting airspace redesigns, standards from the aviation and analytical community, and the opinions of experts we interviewed. FAA's guidance suggests activities for conducting a redesign's operational analysis and establishes specific guidelines for conducting a noise analysis. We found that FAA generally adhered to this guidance, generally following its process for conducting operational analyses and using the noise modeling tool and metric specified in its guidance. In addition, independent experts we interviewed generally agreed that FAA used experienced contractors and used the best available tools to model the operational and noise impacts of the airspace redesign. Furthermore, we found no evidence that the data used are unreliable, and experts interviewed generally agreed that FAA used data that were appropriate and standard to the industry and chose metrics to measure the operational impacts that were reasonable and aligned with the objectives of the airspace redesign. For example, according to FAA and experts we interviewed, such data sources as FAA radar flight track data, U.S. Geological Survey terrain data, and U.S. Census Bureau data are industry standard and generally recognized as providing reliable information. On the basis of generally accepted economic principles and practices, expert opinions, past FAA-sponsored airspace redesigns, and prior GAO reports, we also identified some limitations in FAA's methodology that, if addressed, would provide more information for decision makers and the public. We and the experts we interviewed do not believe these limitations are substantial enough to warrant redoing the analyses, but rather, would help identify ways in which FAA's methodology could be improved in future airspace redesigns. These limitations are not matters that FAA is required by law to address. One limitation is that because FAA assumed that traffic demand and flight operations would not increase in response to airspace system improvements—specifically, delay reductions and operating cost savings—FAA did not account for the potential effect of the system improvements in its operational analysis. FAA also did not fully assess the uncertainty associated with each alternative's estimated impacts: FAA did not consider a range of values for key assumptions and inputs, even though some key inputs, such as the aviation demand forecast, are inherently uncertain. In addition, when evaluating the alternatives, FAA did not analyze economic impacts, such as implementation costs, the effect of the alternatives on the airlines, passengers' time value, or the effect of noise on the quality of life of residents living near the airports. FAA did not estimate the economic impacts of noise in part because the agency is not required under CEQ regulations to do so. We identified two analyses that, while FAA is not required by law or guidance to conduct, could benefit decision makers and the public. First, an uncertainty analysis that examines a range of key assumptions and inputs, such as the aviation demand forecast, would provide additional information on the potential range of the redesign's impact to decision makers and the public. Second, a benefit-cost analysis, which assesses the benefits and costs of each alternative compared to the status quo, would have provided decision makers and the public with more information on whether the estimated benefits of the redesigned airspace justify its costs.

Based on the information available to date, we are unable to fully assess the likelihood that FAA will meet its projected five-year implementation time frame or determine the costs of the airspace redesign project. Although FAA internal guidance recommends that an implementation plan—including specific implementation activities, such as establishing a project schedule or planning for potential challenges—be developed immediately after approval of the airspace redesign, FAA has not developed a detailed implementation plan that outlines these specific activities. However, even though FAA does not have a detailed implementation plan available, FAA began limited implementation at Philadelphia and Newark on December 19, 2007. Without a more detailed implementation plan, we cannot determine whether FAA is likely to meet its projected schedule or how it will deal with potential challenges. In addition, for two principal reasons, the final project configuration and costs are unknown. First, although FAA will use currently available equipment and software, FAA has not determined the type or amount that will be needed to support the common automation platform for air traffic controllers. Second, FAA is currently reviewing whether the common automation platform, which will allow different air traffic controllers to communicate with each other, will be housed in existing FAA facilities, a new facility, or in a consolidated facility. If FAA decides to house this platform in a new facility, FAA officials stated that a final cost estimate for the new facility will not be available until late 2008, and is not projected to be completed until 2015, 3 years after FAA has estimated full implementation of the airspace project. Given that the redesign represents a complex and comprehensive change to the region's airspace, the agency also should consider how it plans to conduct evaluations of the redesign.

One potential strategy for FAA to employ is adaptive management, which is a process that promotes flexible decision making as outcomes from management actions become understood, and is a strategy that GAO has recommended for other federal agencies. Furthermore, adaptive management is recognized by the CEQ and other federal agencies as a useful strategy.

To improve FAA's implementation of the New York/New Jersey/Philadelphia Airspace Redesign, we are recommending that the Secretary of the Department of Transportation direct the Acting Administrator of the Federal Aviation Administration to develop and follow a detailed implementation plan and postimplementation evaluation plan using an adaptive management strategy for this airspace redesign. To improve FAA's effectiveness and accountability in conducting future airspace redesigns we are recommending that the Secretary of the Department of Transportation direct the Acting Administrator of the Federal Aviation Administration conduct uncertainty analyses and benefit-cost analyses when evaluating the potential effect of future airspace redesigns.

In commenting on a draft of this report, DOT stated that it was pleased that the GAO review concluded the actions taken by FAA with regard to NEPA were reasonable. However, DOT also noted that while it did not necessarily agree with everything in the draft report, it is constrained from offering a detailed analysis of the draft report at this time, as the matters the report covers are the subject of pending litigation (see app. V for DOT comments). DOT provided technical comments that we incorporated as appropriate.

BACKGROUND

For several reasons, the New York/New Jersey/Philadelphia airspace is unique and highly complex. First, numerous airports are located within a small geographic area, including some of the nation's most important major airports. In the New York City metropolitan area alone, there are four major airports, including New York's John F. Kennedy International Airport (JFK) and LaGuardia Airport (LaGuardia) and New Jersey's Newark Liberty International Airport (Newark) and Teterboro Airport (Teterboro). Philadelphia International Airport (Philadelphia) is located less than 100 miles southwest of New York City. The region also has numerous satellite airports,[10] which provide substantial aviation service to Connecticut, New Jersey, southeastern New York, and eastern Pennsylvania (see Figure 1). Second, the region's airports provide service to a very large air travel market. Since the New York metropolitan area has one of the largest urban populations in the United States, there is a substantial market for air travel, which is serviced by many domestic and international airlines. Third, three of these airports—JFK, Newark, and Philadelphia—serve as hubs for major U.S. airlines[11] to connect passengers to other flights. Finally, both LaGuardia and JFK have been slot controlled[12] in the past. In January 2007, the authority to issue slot restrictions at LaGuardia and JFK airports expired;[13] however, due to the increased congestion and flight delays at LaGuardia, JFK, and Newark, FAA has imposed temporary slot controls at these airports with the goal of preventing an increase of scheduled flights during peak demand times.

According to FAA, the current airspace structure—which basically has remained the same since the 1960s[14]—is inefficient and cannot meet future travel demands without increasing delays. The airspace was designed for lower air traffic volumes and different types of aircraft, and is characterized by complex, highly interrelated departure and arrival routes. FAA maintains that there is a need to redesign the region's airspace because the current structure cannot efficiently address the increased air traffic levels, the use of new aircraft types, and emerging technologies to control air traffic. In addition, because specific routes used by aircraft taking off from one airport overlap routes from other airports, safety-related inefficiencies ensue. Severe weather conditions also may hinder aircraft's access to departure routes at certain airports. According to FAA, the complexity of the airspace also has contributed to inefficiencies in the air traffic management for the region. To provide air traffic management for the region's airspace, FAA uses a variety of air traffic control facilities and predetermined and coordinated procedures and routes. Each ATC facility has a specific—and different—area of responsibility for the overall management of the regional system. Figure 2 depicts a cross section of the different air traffic control facilities and procedures.

In April 1998, FAA launched the nation's first-ever coordinated, comprehensive National Airspace Redesign program with goals of moving aircraft more efficiently and more safely through the airspace system. The goals of this program are identified in the National Airspace Redesign Strategic Management Plan,[15] and include maintaining system safety, decreasing system delay, increasing system flexibility, increasing system predictability, and increasing user access. The Airspace Management Program Office is responsible for airspace redesign project planning and ensuring that planning goes beyond the design phase and includes an execution and implementation focus. Other organizations also are involved in the national airspace redesign efforts, such as the Environmental Protection Agency, which reviews and publicly comments on the adequacy of FAA's analysis of environmental impacts.

In 1998, as part of its National Airspace Redesign program, FAA initiated the New York/New Jersey/Philadelphia Airspace Redesign. To redesign the airspace, FAA uses its Airspace Management Handbook (Handbook)—which describes step-by-step procedures for airspace design management—to guide the overall management of the redesign project and must follow applicable environmental laws and regulations. The Handbook provides FAA guidance on all phases of the redesign project from characterizing the initial problem to conducting the operational analysis for the airspace redesign to postimplementation evaluation.[16] FAA also is required to comply with NEPA.[17] NEPA requires that a detailed EIS be prepared for all major federal actions significantly affecting the quality of the human environment.[18] To implement NEPA, CEQ has promulgated regulations that set forth specific requirements that all federal agencies must adhere to in the EIS process. Among other things, CEQ regulations require an EIS to (1) specify the purpose of and need for the federal action, (2) describe the environment that will be affected, (3) identify alternatives to the proposed action and identify the agency's preferred alternative(s), (4) present the environmental impacts of the proposed action (including the direct and indirect effects and cumulative impacts), (5) identify any adverse environmental impacts that cannot be avoided should the proposed action be implemented, and (6) identify any irreversible and irretrievable commitment of resources that would occur should the proposed action be implemented. Agencies also must make diligent efforts to involve the public throughout the EIS process.[19] For additional direction on implementing NEPA and CEQ regulations, FAA developed Orders[20] that govern, among other things, its EIS process. In addition, Executive Order 12898,[21] accompanying CEQ guidance, and DOT Order 5610.2[22] set forth environmental justice directives, which FAA must adhere to in conducting its environmental review.

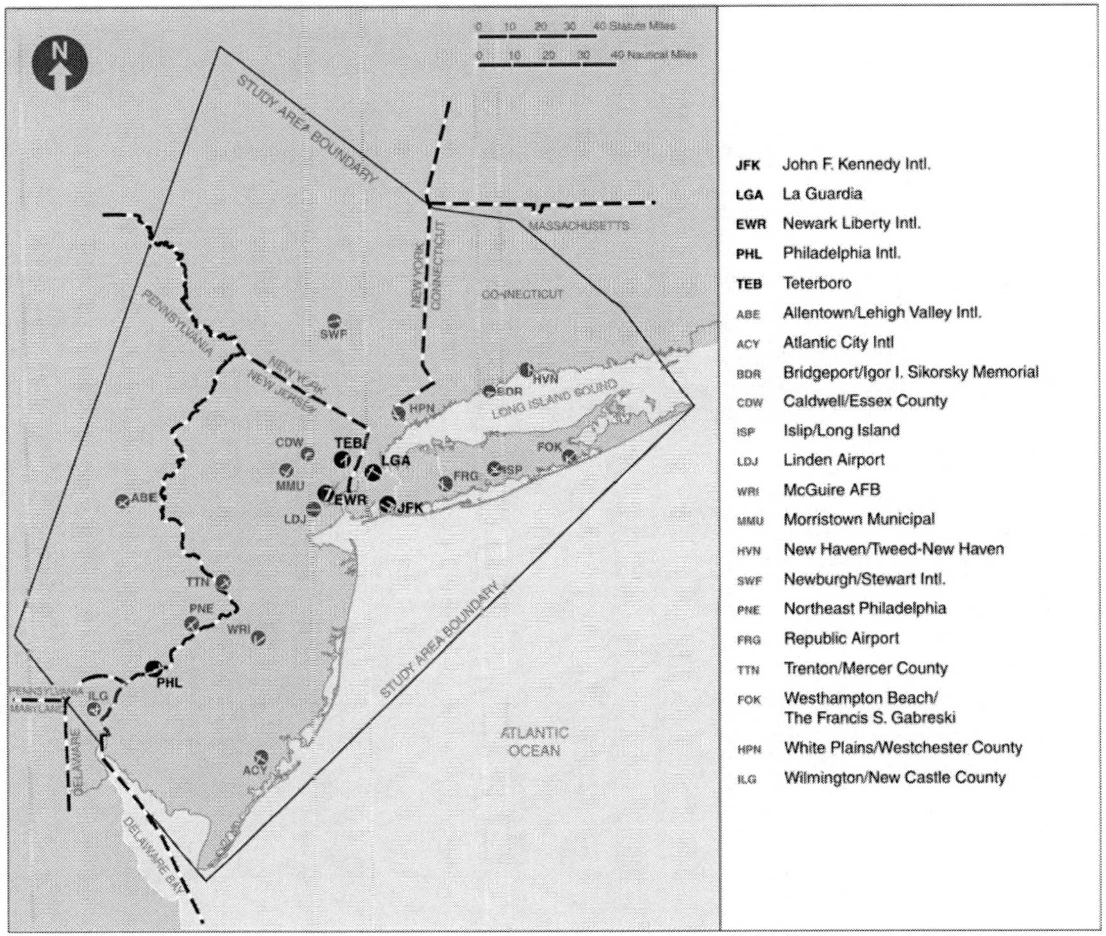

Figure 1. Map of Regions with Major Airports and Satellite Airports Included in FAA's Analysis

Source: FAA.

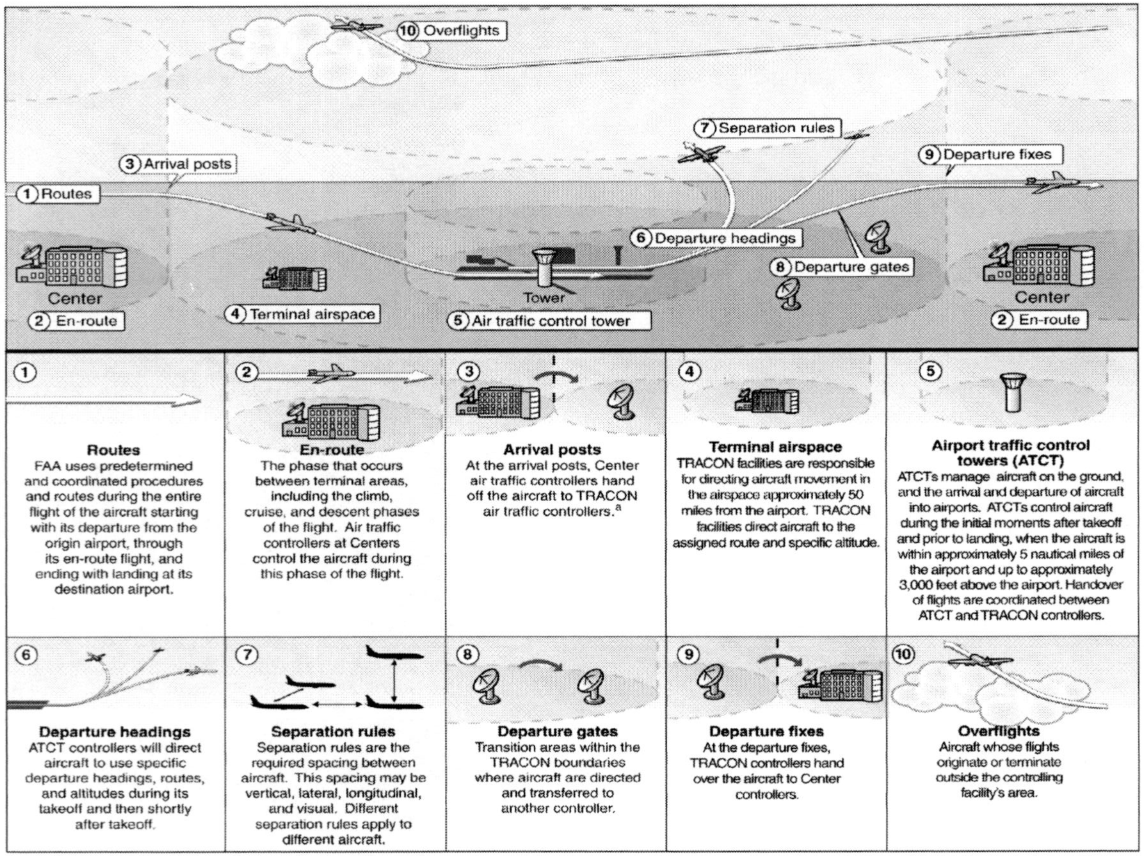

Figure 2. General Representation of a Cross Section of the Different Air Traffic Control Facilities and Procedures

Source: GAO (data); Art Explosion (images).

[a]Terminal Radar Approach Control (TRACON)—An FAA air traffic control facility that uses radar and two-way radio communication to provide separation of air traffic within a specific geographic area in the vicinity of one or more large airports.

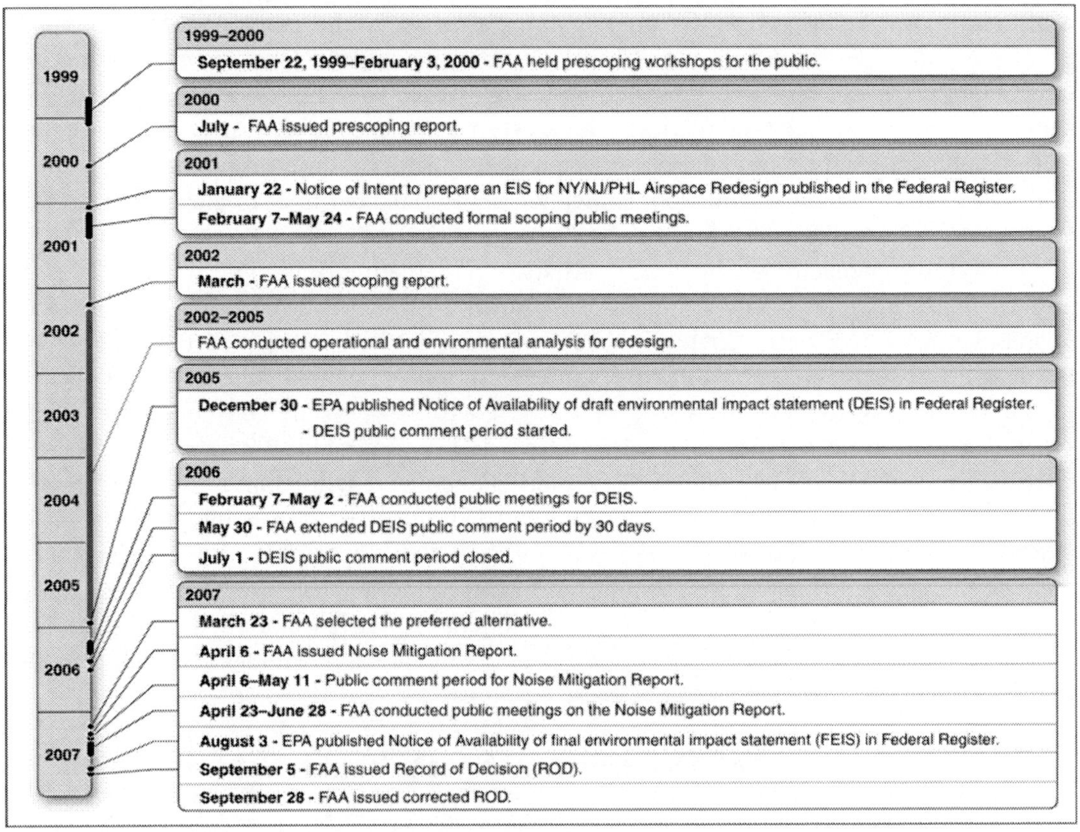

Figure 3. Timeline of FAA's New York/New Jersey/Philadelphia Airspace Redesign Project

Sources: GAO and FAA.

The executive order directs agencies to address "disproportionately high and adverse human health or environmental effects of [an agency's] programs, policies, and activities on minority populations and low-income populations…."[23]

Figure 3 shows some of the key dates in the preparation of FAA's airspace redesign and the EIS for the New York/New Jersey/Philadelphia project.

After determining that it would prepare an EIS, and following 5 years of scoping and options development, in December 2005, FAA issued a draft EIS for public comment that identified four basic alternatives. The first alternative, mandated by NEPA as a baseline, was the Future No Action Alternative, which represents what the airspace structure would be if the airspace redesign did not occur. The second alternative, the Modifications to Existing Airspace Alternative, would have made only slight adjustments within the existing airspace. That is, this alternative would have maintained the current boundaries of the control tower, the Terminal Radar Approach Control (TRACON) facility, and the Center facility. The third alternative, the Ocean Routing Airspace Alternative, would have sent all departing flights from Newark over the Raritan Bay to the Atlantic Ocean before turning them back over land to head to their departure gates. The main purpose of this alternative was to reduce the noise impacts on the citizens of New Jersey. The final alternative was evaluated using two versions. The Integrated Airspace Alternative Variation without Integrated Control Complex (ICC) would have changed the current boundaries of the airspace, to the extent possible, but would have done so with the existing ATC facilities and equipment. The second version, the Integrated Airspace Alternative Variation with ICC, would have fully integrated the airspace using a common automation platform for ATC equipment. It also may have involved consolidating air traffic controllers into a consolidated facility.[24] Figure 4 provides a summary of the different alternatives, the eight evaluation criteria applied to each alternative, and the measures or "metrics" used to compare each alternative.

FAA recommended the Integrated Airspace Alternative Variation with ICC as its preferred alternative, as it scored highest on 10 of the 13 metrics FAA used to compare the alternatives. Appendix III provides the full operational comparison of the alternatives. Because the preferred alternative involves integrating the New York TRACON and portions of the surrounding Centers' airspace, air traffic controllers will be able to reduce aircraft separation rules from 5 miles to 3 miles over a larger geographical area than the current airspace structure allows. In addition, the preferred alternative will modify other key airspace components. Appendix IV provides a more detailed summary of the Integrated Airspace Alternative Variation with ICC. Under this alternative, for example, FAA will be able to increase the number of departure headings air traffic controllers can assign to aircraft during take-offs, and adjust the routes air traffic controllers can assign aircraft during their final approach to an airport. Figure 5 shows an example of departures without and with dispersal headings (increasing the number of departure headings).

After the preferred alternative was announced, FAA initiated a noise mitigation study to develop measures to alleviate, to the extent possible, the noise impacts associated with the preferred alternative. The study identified several mitigation measures for analysis—some of which were later incorporated into the preferred alternative based on the results of this analysis. The mitigation measures changed some aspects of the preferred alternative, including reducing the number of departure dispersal headings, shifting certain departure routes, and changing certain arrival altitudes. According to FAA, the mitigation measures incorporated into the preferred alternative would not substantially reduce the operational gains that FAA believes would occur once the mitigated preferred alternative is fully implemented. For example, at Philadelphia, FAA originally planned to increase the number of new departure dispersal headings on one of its runways from one to six, which would have allowed air traffic controllers to decrease the time between aircrafts on take-offs and increase the number of departures in a given amount of time. However, FAA's analysis of the mitigation measures recommended that the number of headings could be reduced from six to three headings, while minimizing the loss of operational efficiency. The preferred alternative with mitigation measures was identified as the selected project in FAA's Record of Decision, which was issued on September 5, 2007.[25]

FAA officials have projected a 5-year time frame for full implementation of the redesigned airspace, which will be implemented in four stages. According to FAA, each stage will take 12 to 18 months to implement; elements that do not require large-scale changes to the current airspace structure will be implemented first, while more complex changes will be implemented later. For example, during the first stage, implementation focuses on procedural changes, such as adjusting the departure dispersal headings at three of the major airports. FAA implemented the use of additional departure dispersal headings at Philadelphia and Newark as of December 19, 2007.

Alternatives	
Future No Action Airspace Alternative	Maintains the current airspace structure
Modification to Existing Airspace Alternative	Modifies existing airspace but maintains the current boundaries of the ATC facilities
Ocean Routing Airspace Alternative	Sends all Newark departing flights over the Raritan Bay to the Atlantic Ocean before routing flights back over land to head to their departure gates
Integrated Airspace Alternative Variation without Integrated Control Complex	Combines the New York TRACON airspace with portions of surrounding Centers' airspace to permit more seamless operations without a common automation platform or a consolidated facility
Integrated Airspace Alternative Variation with Integrated Control Complex	Combines the New York TRACON airspace with portions of surrounding Centers' airspace to permit more seamless operations with a common automation platform and the possibility of a consolidated facility

Evaluation Criteria	How Measured
Reduce complexity	• **Jet route delays plus time below 18,000 feet:** Measures over a 24-hour period average flight delay and average flight time for a departing aircraft to reach 18,000 feet altitude. • **Arrival distance below 18,000 feet:** Measures average distance traveled by an arriving aircraft from 18,000 feet altitude to landing.
Reduce voice communication	• **Maximum interfacility handoffs per hour:** Measures number of controller-to-controller communications an hour to transfer aircraft responsibility from one ATC facility to another.
Reduce delay	• **Traffic weighted arrival delay 2011 and traffic weighted departure delay:** Measures the difference in estimated time required for an aircraft to arrive or depart absent any delays and to arrive or depart under a specific alternative.
Balance controller workload	• **Equity of west gate fix traffic counts:** Measures the balance of aircraft traffic among departure fixes—where aircraft responsibility is handed over between ATC towers—in a particular departure gate—the transition area within TRACON boundaries.
Meet system demands and improve user access to system	• **End of day's last arrival push:** Measures time when the final bank of scheduled flights from all study area airports enters the TRACON system.
Expedite arrivals and departures	• **Time below 18,000 feet:** Measures average flight time spent descending or climbing per flight in a 24-hour period. • **Change in route length per flight:** Measures the difference in distance flown between Future No Action Airspace Alternative and each alternative. • **Change in block time:** Measures average departure gate to arrival gate flight time in a 24-hour period.
Flexibility in routing	• **Minutes of delay saved per flight per day:** Measures modeled delay caused by waiting out or flying around a 4-hour weather disruption divided by the total number of flights.
Maintain airport throughput	• **Arrival maximum sustainable throughputs:** Measures number of landings per hour. • **Departure maximum sustainable throughputs:** Measures number of takeoffs per hour.

Figure 4. Alternatives Considered under the FAA New York/New Jersey/Philadelphia Airspace Redesign EIS
Sources: GAO and FAA.

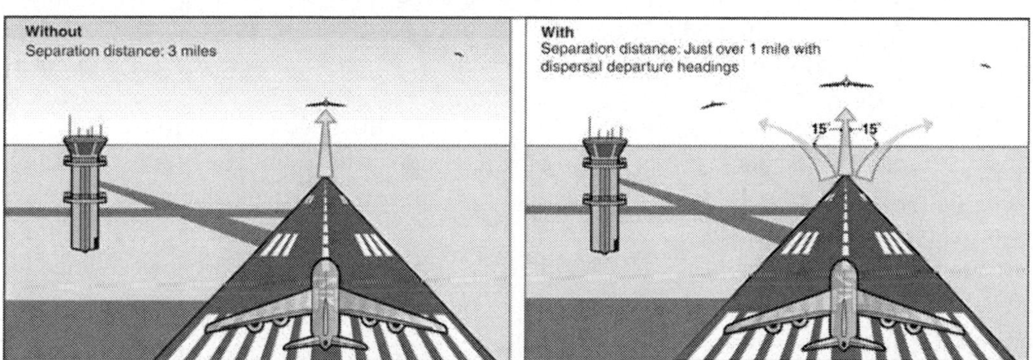

Figure 5. Example of Departures without and with Dispersal Headings
Source: GAO (data); Art Explosion (images).
Note: These separation distances are based on normal conditions with similar types of aircraft. Separation distances can be increased depending on certain weather conditions or if smaller aircraft are taking off after a large-body aircraft.

FAA Complied with Key Legal Requirements in Conducting Its Environmental Review for the Regional Airspace Redesign

Our evaluation of FAA's compliance with NEPA and related requirements, in conducting the airspace redesign, was based on established court precedent applying NEPA and its implementing regulations. In deciding whether an agency adequately carried out these requirements, a court uses a standard of review articulated in the APA, which is deferential to agency decision making—whether an agency's actions were "arbitrary, capricious, an abuse of discretion, or otherwise not in accordance with law."[26] Courts interpret this APA standard as requiring that an agency act reasonably in carrying out NEPA's requirements and that the agency's ultimate decisions be reasonable and not arbitrary and capricious. We reviewed FAA's compliance with respect to five key issues: the statement of the project's purpose and need, evaluation of alternatives, consideration of the project's environmental effects, public involvement, and environmental justice matters. We selected these five issues based on public concerns raised during and after the EIS process, congressional interest, the views of experts we interviewed, and GAO's evaluation of the range of concerns presented. Applying these NEPA requirements and the APA's reasonableness standard, we concluded that FAA complied with respect to these five key issues. Our detailed legal analysis is contained in appendix II and is summarized below.

FAA's Statement of Purpose and Need Complied with Requirements

FAA's statement of purpose and need, which defined the objectives and parameters of the project, complied with NEPA requirements. FAA developed the following purpose and need statement:

> "The purpose of the airspace redesign is to increase the efficiency and reliability of the airspace structure and ATC system. The need is to accommodate growth in aircraft operations while maintaining safety, mitigating delays, and accommodating changes in the types of aircraft using the system."[27]

CEQ regulations require that FAA "...briefly specify the underlying purpose and need to which the agency is responding in proposing the alternatives including the proposed action."[28] Additionally, FAA Order 1050.1E requires that the purpose and need statement "...present[] the problem being addressed ... and essentially provide [t]he parameters for defining a reasonable range of alternatives to be considered."[29] FAA specified the purpose and need and defined the parameters of the project such that a reasonable number of alternatives could be developed—the statement was not too narrowly or too broadly defined. During the EIS process, members of the public and certain advocacy groups supported the inclusion of noise minimization in the redesign's stated purpose. According to FAA officials, although the agency was aware that noise was an important issue to the public from the onset of the EIS process, it is not FAA's policy to minimize noise impacts for one community at the expense of another. FAA noted in the final EIS that its highest priority must be to maintain a safe and secure airspace. Based on our legal review, we find FAA's decision to exclude noise from the purpose and need to be reasonable, in part, because NEPA does not require agencies to elevate environmental concerns—in this case noise—over other appropriate considerations. Rather, NEPA requires only that agencies consider the environmental impacts of their actions as part of their decision making. FAA's decision also is consistent with court precedent, including a court decision upholding an FAA purpose and need statement excluding noise from a major airport expansion, even though the project would have caused significant noise impacts.[30]

FAA's Range and Evaluation of Alternatives Complied with Requirements

FAA complied with the various NEPA requirements for evaluating the four alternatives, including the selected project, during the EIS process. In the EIS, the agency included a no action alternative and discussed alternatives eliminated from detailed analysis. In addition, the evaluated alternatives were responsive to the agency's purpose and need and were reasonably developed, rigorously explored, and objectively compared to each other throughout the EIS.

- *Representation of a baseline of current airspace conditions in an alternative. By including a Future No Action Alternative, FAA complied with the requirement[31] to provide a no action alternative that provided a baseline of current conditions to which the other alternatives were compared.*

- *Discussion of alternatives eliminated. FAA "...briefly discuss[ed] the reasons..."[32] for alternatives that were evaluated but subsequently eliminated from detailed consideration in the EIS.[33] For example, during the formal scoping period, there was public support for the development of alternatives that included either alternate modes of transport or improvements to airport infrastructure (such as additional runways at Philadelphia) as means to reduce delays or improve operational efficiency. In addition, during subsequent public comment periods stakeholders suggested other alternatives, such as peak hour demand controls. Within the EIS, FAA explained that while it had considered these alternatives, it ultimately eliminated them because they did not meet the purpose of reducing the inefficiencies of the airspace.*

- *Assessment and evaluation of reasonable alternatives. In order to be "reasonable," the range of alternatives must first be responsive to the purpose and need.[34] Our review showed that to develop the alternatives, FAA worked with modelers and air traffic controllers in an iterative and collaborative process specifically to design alternatives that were feasible within the parameters of the redesign's purpose and need. Furthermore, FAA used the project's purpose and need to evaluate the alternatives, both quantitatively and qualitatively, as each evaluative criterion informed an aspect of the stated purpose and need. Some community stakeholders and government officials criticized the range of alternatives considered, pointing out that FAA did not include an alternative that focused on noise reduction. However, FAA officials explained that as noise reduction was not among the defined purposes of the redesign, it did not have to be considered when developing the alternatives. Furthermore, with the inclusion of the noise mitigation for the preferred alternative, FAA believes it minimized environmental impacts, such as noise, to the extent possible while continuing to address the purpose and need. Despite stakeholder concerns, as noted above, we find FAA was responsive to the project's purpose and need, which did not include noise reduction; additionally, the selected project, which did include noise mitigation measures, minimized noise impacts to the extent possible while still addressing the project's purpose and need.*[35]

- *Alternatives' comparative merits. To comply with the CEQ regulation, alternatives also must be discussed in reasonable detail;*[36] *there must be a rigorous exploration of each alternative.*[37] *As demonstrated by the thorough discussion throughout the EIS, FAA substantially evaluated each of the four primary alternatives, including both operationally—discussing routing procedures, reduction of complexity, delay, voice communication, and balancing controller workload—and in terms of the alternative's environmental consequences. FAA also compared the alternatives using specified criteria and presented the alternatives in summary comparison tables, thereby enabling "reviewers [to] evaluate [the alternatives'] comparative merits."*[38]

FAA's Decision Not to Consider Environmental Effects of the Potential Growth Inducement Resulting from Airspace Redesign Complied with Requirements

FAA complied with NEPA in its decision not to analyze the environmental effects of a potential increase in capacity resulting from the regional airspace redesign. CEQ regulations require that within the EIS process, an agency must consider the direct and indirect effects and cumulative impacts of a proposed action when evaluating the environmental consequences.[39] Indirect effects,[40] which are reasonably foreseeable future effects caused by the current project, may include growth-inducing effects, among others. FAA did not analyze the indirect effects of potential growth resulting from the redesign because it found that the redesign would not in itself create a growth-inducing effect or increase capacity. Therefore, FAA concluded that such an analysis was not required. Members of the public and government agencies criticized FAA's decision, stating that the efficiencies gained from a redesigned airspace would lead to an increase in capacity and consequently would increase air traffic and emissions. In other words, an indirect effect would occur as a result of the redesigned airspace.

We found that FAA's decision not to consider these potential indirect environmental effects was reasonable under court decisions applying these NEPA requirements. In the aviation context, courts have uniformly upheld similar decisions by FAA not to analyze the effects of induced growth where the purpose of the project was not growth inducing and did not add runway capacity. The U.S. Court of Appeals for the Ninth Circuit, for example, ruled that FAA complied with NEPA even though it did not consider the possible growth-inducing effects of altering arrival procedures at Los Angeles International Airport, where there would be no changes to the airport's infrastructure. The court explained that although growth might certainly be foreseeable, "the project was implemented in order to deal with existing problems; the fact that it might also facilitate further growth is insufficient to constitute a growth-inducing impact…" under CEQ regulations.[41] Likewise in this case, the purpose of the regional airspace redesign was to increase the efficiency and reliability of the existing airspace structure and ATC system—not growth inducement—and the project does not add runway capacity. Therefore, we found that FAA was not required to analyze potential environmental effects of increased growth that might result from the airspace redesign.

FAA Complied with Public Participation Requirements

FAA complied with NEPA requirements to make "…diligent efforts to involve the public…"[42] during the EIS process by (1) conducting an early and open process to determine the scope of issues,[43] (2) providing notice of and holding public meetings,[44] and (3) soliciting and responding to public comments.[45] We determined that FAA used an "early and open" process, as demonstrated by an extensive prescoping campaign: 31 "prescoping" workshops were conducted prior to the formal scoping period[46] which drew 1,174 attendees and prompted 712 written comments. As with the scoping meetings, according to FAA, the purpose of these meetings included improving public understanding of the project and increasing FAA's understanding of public issues that would need to be addressed as part of the redesign process. For these meetings, as well all other public meetings conducted for the redesign, FAA advertised in local media outlets for upcoming meetings and developed a mailing list to send notifications about public meetings and periodic updates on the redesign.[47] FAA continued to conduct public meetings and to solicit public comments for the scoping and draft EIS periods by holding 58 public meetings that drew over 2,000 attendees. In addition, FAA held seven public meetings during the Noise Mitigation Report comment period, drawing the largest number of participants—2,200. According to FAA officials, meeting locations for the different phases of the EIS process were selected based on multiple criteria including site suggestions from government officials and their staffs, proximity to areas that would be by the proposed action, and FAA's goal to hold at least one meeting in each affected state.[48]

During the EIS process, public comments included criticism that the meeting locations were not widely dispersed around the affected states. There also was criticism that the comment period for the Noise Mitigation Report was too short and that there were too few public meetings regarding this report, which addressed one of the most controversial and technically complex issues of the redesign. As demonstrated in table 1, although FAA held the fewest number of public meetings for this phase of the EIS process, the meetings had the highest number of attendees. FAA officials stated that they recognized that noise mitigation issues would attract a lot of public attention. Despite stakeholder concerns, we concluded that FAA complied with their requirements.[49]

FAA also solicited public comments and considered and responded to comments, as required by CEQ regulations, in an adequate manner.[50] The public comment period for the Scoping and Noise Mitigation periods lasted at least 30 days. For the draft EIS comment period, FAA extended the initial 5-month comment period by 30 days to allow the public more time to submit its comments. FAA also solicited public comments at the public meetings and accepted them during the public comment periods via U.S. mail and e-mail. Receiving numerous public comments, FAA categorized and responded to all substantive comments for the draft EIS comment period in the final EIS. Based on FAA's solicitation and response to public comments received throughout the EIS process, we concluded FAA's actions complied with requirements. See table 1 for a summary of public participation opportunities during the EIS process.

Table 1. Summary of Public Participation Opportunities during the EIS Process

Phase	Total number of public meetings	Time period for public meetings (days)	Total number of public attendees	Length of comment period (days)
Prescoping	31	135	1,174	
Scoping	28	107	1,031	159
Draft EIS	30	85	1,166	184
Noise Mitigation	7	67	2,200[a]	36

Sources: GAO and FAA.

[a]The source documentation cites that "over 2,200 people" attended the noise mitigation meetings. As a specific number was not included by FAA in its documentation, GAO used the number that was listed—2,200.

FAA Complied with Environmental Justice Directives

FAA complied with environmental justice directives set forth in Executive Order 12898,[51] accompanying guidance,[52] and DOT Order 5610.2,[53] including (1) identifying and addressing, as appropriate, disproportionately high and adverse environmental impacts of the proposed action on minority and low-income populations and (2) eliciting public involvement from minority and low-income populations during the EIS process. Understanding that potential significant noise impacts could result from the alternatives, FAA identified and addressed any disproportionately high and adverse human health or environmental impacts on minority and low-income populations in the study area. First, FAA used U.S. Census Bureau block data to identify the areas that could be significantly impacted by noise. Then, FAA determined whether the impact on minority and low-income populations was disproportionate.[54] While certain areas were identified as locations with significant environmental impacts to minority and low-income populations, the mitigation measures outlined in the selected project reduced the expected impacts by 2011 to below significant levels, as defined by FAA's standard of noise exposure. Therefore, FAA concluded that no additional measures were needed to minimize noise impacts for the identified minority and low-income populations. By conducting this assessment, FAA met the directive to identify and address disproportionate significant impacts on low-income and minority populations. [55]

To comply with DOT Order 5610.2 and guidance accompanying Executive Order 12898, FAA elicited opportunities for public involvement of minority and low-income populations for the public meeting periods of the EIS. As the final EIS described, prior to the start of the public meeting periods, the agency worked with congressional offices to determine appropriate locations that would accommodate the specific needs of minority and low-income populations. According to FAA officials, a number of meetings during each public meeting period of the EIS process were held in locations accessible by public transit. Additionally, public meetings throughout the EIS process period were held in low-income and minority communities. Foreign language translators were also provided at certain meetings and information was presented in a variety of formats. Lastly, FAA published advertisements for the draft EIS public meetings in specialized local foreign language media throughout the redesign's affected states. Thus, FAA complied with requirements outlined in Executive Order 12898 and related guidance and Orders.

Methodology Used to Assess Key Impacts Was Reasonable, However Additional Analyses Could Have Benefited Decision Making and the Public's Understanding

FAA's methodology to assess the operational and noise impacts was reasonable, based on FAA's guidance for conducting airspace redesigns, standards from the aviation and analytical community, and opinions of experts we interviewed. We used FAA's guidance—specifically the Handbook and FAA Order 1050.1E—to establish criteria to examine FAA's methodology. The Handbook suggests activities for conducting a redesign's operational analysis. In addition, FAA Order 1050.1E establishes specific guidelines for conducting its noise analysis. We found that FAA generally adhered to its guidance in conducting the redesign. Specifically, FAA generally followed the process outlined in its Handbook and used the noise modeling tool and metrics specified in FAA Order 1050.1E. In addition, experts we interviewed generally agreed that FAA used experienced contractors, used best available tools to model the operational and noise impacts of the airspace redesign, used data that were appropriate and standard to the industry, and chose metrics to measure the operational impacts that were reasonable and aligned with the objectives of the airspace redesign. However, based on generally accepted economic principles and practices, expert opinions, past FAA-sponsored airspace redesigns, and prior GAO reports[56] we identified four limitations in FAA's operational and noise analyses and an additional limitation in not analyzing economic impacts when evaluating the alternatives. Based on our analysis and the opinions of experts we interviewed, we do not believe that these limitations are substantial enough to warrant redoing the analyses, but help identify ways in which FAA's methodology could be improved in future airspace redesigns. Consequently, we identified two types of analyses—uncertainty analyses and benefit-cost analyses—that, although not required by law, would provide more comprehensive information for decision makers and may increase public understanding of FAA's process and decision making in future airspace redesigns.

Methodology Used to Assess Operational and Noise Impacts Was Reasonable

FAA's methodology to assess the operational and noise impacts was reasonable, based on FAA's guidance for conducting airspace redesigns, standards from the aviation and analytical community, and opinions of experts we interviewed. To examine FAA's methodology, we focused on five components—process, contractors, modeling tools, data, and metrics—which we identified from FAA's guidance, interviews with FAA officials and experts, and prior GAO reports.[57] Specifically, we found that FAA generally followed the process outlined in its Handbook and used the noise modeling tool and metric specified in FAA Order 1050.1E. In addition, FAA used experienced contractors and, according to experts we interviewed, the best available modeling tools in its operational and noise analyses. Furthermore, we found no evidence that the data used are unreliable, and experts interviewed generally agreed that FAA used data that were appropriate and standard to the industry and chose metrics to measure the operational impacts that were reasonable and aligned with the objectives of the airspace redesign.

Process

FAA generally followed the process outlined in its Handbook. Airspace studies are generally iterative and involve substantial professional judgment. FAA formulated alternatives for redesigning the airspace through an iterative process with FAA's airspace redesign team[58] and MITRE (the organization responsible for the operational analysis). Initially, as outlined in FAA's Handbook, conceptual ideas were developed based on the project's objectives. Then using an operational modeling tool, the conceptual ideas were simulated to check that they were technically feasible. If there were parts of the conceptual design that did not work as evidenced through the modeling, the design was adjusted, and the new design was simulated. This iterative process was used in designing the Modifications to Existing Airspace Alternative and the two variations of the Integrated Airspace Alternative.[59]

Once the conceptual alternatives were designed, FAA collaborated with MITRE to develop simulation models for each alternative to measure the operational impacts. Performance metrics,[60] which are selected based on the project's objectives, were used to measure the operational impacts, or system improvements, of each alternative. As discussed in the Handbook, the purpose of the metrics is to allow FAA to evaluate and compare the alternatives, which contributed directly to FAA's decision. The Total Airspace and Airport Modeler (TAAM) was the primary tool used to model the alternatives' operations and quantify the operational impacts in terms of the performance metrics. Additional modeling tools were used to supplement the operational analysis. FAA, MITRE, and air traffic controllers validated each operational design through an iterative process, as outlined in the Handbook, to ensure that the models would provide reliable results by reviewing the model output, making adjustments, and continuing that process until there was agreement that the design was valid.

After the operational designs were validated, the four alternatives were evaluated for environmental impacts, and the results were included in the draft EIS. In conducting the noise analysis, FAA collaborated with its environmental contractors and MITRE to develop scenarios for each alternative to measure the noise impacts in terms of a cumulative noise metric. The Noise Integrated Routing System (NIRS) model was the primary tool used to measure the noise impacts. While developing the input for the noise model, quality assurance checks were conducted by the modelers to help assure accuracy in the model output. Using an iterative process, FAA's airspace redesign team, the environmental contractors, and MITRE reviewed portions of the input for the noise model and output from the operational simulations and validated them against one another to help ensure consistency between the operational and noise analyses.

FAA selected the Integrated Airspace Alternative Variation with ICC as its preferred alternative because, according to FAA, it best met the project's objectives. However, after selecting the preferred alternative, FAA began the process of identifying measures to mitigate the noise impacts associated with the preferred alternative. This process again was iterative and involved identifying potential noise mitigation strategies and using operational and noise modeling tools to measure the operational and noise impacts. The potential noise mitigation measures were continually adjusted and evaluated until measures were identified that would reduce noise impacts without substantial adverse impact on operational efficiency. Several noise mitigation measures were incorporated into the preferred alternative, resulting in the mitigated preferred alternative—the Integrated Airspace Alternative Variation with ICC and mitigation measures. In the Record of Decision, FAA identified this alternative as its selected project.

The operational and noise models required substantial amounts of data. As highlighted in the Handbook, some data were used as direct inputs to the models, such as the airport and runway configurations for the airports, while other data were used as supporting data when developing the model inputs. For example, in the noise analysis, a sample of radar data was analyzed to develop the locations of the terminal and en route flight tracks. In addition, numerous assumptions were required for the operational and noise analyses. FAA and its contractors made assumptions not only when developing the simulations (including the simulations' scope and detail), but also when developing the data inputs, including the aviation demand forecast. For example, the operational models included details down to the airport level based on the judgment that the redesign has the possibility of improving flexibility of runway use. In addition, while developing the aviation demand forecast, which is a key input to the models, numerous assumptions were made, including assumptions about future traffic demand and the introduction of new aircraft. Figure 6 represents a simplified representation of FAA's methodology used to assess operational and noise impacts.

Contractors

Key to our finding that FAA's contractors were reasonable is the contractors' substantial experience conducting FAA-sponsored aviation studies. Experience is important because both operational and noise analyses inherently involve substantial professional judgment, including determining the detail included in the model to validating the model and interpreting the results. For example, the Handbook states that some of the key consequences of airspace changes may not be captured well by models and that a model's quantitative metrics often need to be supplemented with a qualitative review by knowledgeable experts. For this reason, one expert we interviewed stated that the results of a study depend on the contractor's expertise. MITRE, the organization that conducted the operational analysis, is well-respected in the industry, according to experts we interviewed. MITRE's aviation program has a long-standing relationship with FAA, as evidenced by its designation as a Federally Funded Research and Development Center, and according to MITRE representatives, has generally conducted FAA's more complex airspace redesigns. One expert highlighted that over the years, MITRE has improved the main modeling tool used in the operations analysis to create more valid simulations. Metron Aviation, Inc., a contractor responsible for about half of the noise modeling, developed the NIRS model for FAA and has worked on several airspace studies, including the Potomac Consolidated TRACON Facility Airspace Redesign and the Chicago Terminal Airspace Project. Landrum & Brown and Northrop Grumman, the contractors responsible for the other half of the noise modeling, also have experience with airspace projects.

Modeling tools

The experts we interviewed agreed that FAA used generally accepted tools to model the operational and noise impacts of the airspace redesign. TAAM, the primary operations modeling tool, is used to model airspace, airports, and traffic flows. Some experts said that TAAM was generally the best tool to use for this airspace redesign. It allows for fast-time simulation and analysis of changes to route or sector structure, procedures, and traffic levels. For that reason, FAA and aviation stakeholders, such as airport owners, also have used TAAM to model airport and airspace operations.

Noise analysis experts that we interviewed uniformly agreed that the NIRS modeling tool was the best available tool for this multiairport airspace redesign. NIRS is capable of evaluating complex air traffic designs involving high-altitude routing, broad-area airspace changes affecting multiple airports, and other airspace modifications in the terminal and en route environments. While this airspace redesign presented noise exposure in terms of 1 metric, NIRS is capable of computing 13 predefined noise metrics. NIRS was first released as a prototype model in 1998 and has been used in two previous FAA airspace studies (the Potomac Consolidated TRACON Facility Airspace Redesign and the Chicago Terminal Airspace Project). In addition, FAA Order 1050.1E states that the NIRS model must be used for noise analysis of regional airspace studies.

Data

Based on our examination of the data sources and steps taken by FAA and its contractors to ensure the completeness and accuracy of the data, we found that data used in this airspace redesign (such as the airport runway configurations and flight tracks) were appropriate and standard to the industry. While we did not conduct our own reliability assessments for any data used by FAA, our review disclosed no evidence that these data are unreliable. According to FAA and experts we interviewed, the data sources are industry-standard and generally recognized as providing reliable information. For example, most of the data were obtained from governmental sources, such as the radar flight track data, U.S. Geological Survey terrain data, and U.S. Census Bureau data. Data obtained from nongovernmental sources (such as the Official Airline Guide, Airframe Manufacturers Forecast, and other aviation and population forecasting studies) were primarily used to support aviation demand forecasting, rather than as direct inputs for modeling. While FAA did not employ specific procedures to test these data's reliability, according to FAA and its contractors, basic professional care was undertaken. After data were obtained, generally in electronic format to avoid data entry errors, FAA and its contractors took additional steps to ensure that the data were complete and accurate. For example, FAA and its contractors conducted consistency checks throughout the analyses. According to FAA, consistency checks were used to identify discrepancies between data sets that, if identified, were then reviewed and rectified. In addition, consistency checks were conducted on model inputs that were developed. For example, in the noise analysis, a sample of radar flight track data was analyzed to develop the locations of the terminal and en route flight tracks and the flight track dispersions,[61] which were used as direct inputs to the noise model. Furthermore, two experts we interviewed noted that the use of actual radar data in the noise analysis was a good practice.

Metrics

FAA generally used appropriate metrics to measure the operational noise impacts. According to the Handbook, the selection of performance metrics should be linked to development of the alternatives, as well as the project's objectives. For this airspace redesign, FAA used several performance metrics from the Handbook's list of standard performance metrics. For example, the Change in Route Length per Flight metric is correlated to the Handbook's Average Arrival Flight Distance metric. The Handbook states, however, that the standard performance metrics do not preclude the use of additional metrics. For this airspace redesign, FAA also selected and customized additional performance metrics based on the airspace redesign's specific goals of improving user access and increasing flexibility. Experts that we interviewed generally agreed that the performance metrics used to measure the operational impacts were reasonable and aligned with the project's objectives. One expert also said that customizing performance metrics was a good practice because the effects of the changes made in the airspace redesign could then be isolated.

To measure noise exposure, FAA used the Day-Night Average Sound Level (DNL) metric that, under FAA Order 1050.1E, must be used as the primary noise metric.[62] DNL averages the magnitude of sound levels generated by all individual events occurring during a 24-hour period, with a 10-decibel penalty for noise events occurring during typical sleeping hours (between 10:00 p.m. and 7:00 a.m.). To assess the significance of the noise impacts, FAA used the generally accepted method, which sets a significant noise impact threshold criteria based on a dose-response relationship—the correlation of DNL to the percentage of population highly annoyed by recurring noise sound events. The correlation shows that the percentage of people highly annoyed by noise exposure increases significantly above DNL of 65 decibels. Some aviation noise consultants, however, have noted disadvantages of DNL as a metric to measure noise, and suggest the use of supplemental metrics that measure the frequency and intensity of individual noise events, which may provide a more adequate measure of aviation noise as experienced by people. In 1992, after a comprehensive review of measurement approaches, the Federal Interagency Committee on Noise (FICON) concluded there were no other metrics of sufficient scientific standing to replace DNL, and consequently DNL is used by all federal agencies when analyzing airport-related noise in environmental assessments and impact statements.[63] This conclusion was still valid as of 2000, according to the chairman of the Federal Interagency Committee on Aviation Noise,[64] which focuses on aviation research related to noise. Although FICON recommended in 1992 continuing the use of the DNL noise metric as the principal means of describing airport noise exposure, the Federal Interagency Committee on Aviation Noise recognized in 2002 that supplementing this description with noise analyses based on alternative metrics would provide valuable information that is not easily captured by DNL, and that the use of some supplemental noise metrics can provide a more meaningful estimate of impacts than a single DNL estimate.

Our Review Identified Limitations, Which Had They been Addressed, Would Have Provided More Comprehensive Information for Decision Makers and the Public

Based on generally accepted economic principles and practices, expert opinions, past FAA-sponsored airspace redesigns, and prior GAO reports,[65] we identified four limitations in FAA's operational and noise analyses and an additional limitation in not analyzing the economic impacts when evaluating the alternatives. We and the experts we interviewed do not believe that the limitations identified are substantial enough to warrant redoing the analyses. Addressing these limitations would have provided more comprehensive information for decision makers and may have increased public understanding of the process and decision. For example, although not required, FAA did not fully assess the uncertainty associated with its estimated impacts—that is, it did not consider a range of values for key assumptions and inputs—even though some key inputs, such as the aviation demand forecast, are inherently uncertain. In addition, although FAA was not required to analyze the economic impacts (such as implementation costs or the effect of noise on the quality of life of residents living near the airports) when evaluating the alternatives, GAO has previously highlighted the importance of analyzing economic impacts for agency decision making.

FAA did not Account for the Potential Effect of Delay Reductions and Operating Costs Savings on Passengers and Airline Traffic

Because FAA assumed that traffic demand and flight operations would not increase in response to airspace system improvements—specifically, delay reductions and operating cost savings—FAA did not account for the potential effect of system improvements in its operational analysis. While our legal review found that FAA's decision to not analyze the potential increase in capacity in its EIS is reasonable from a legal standpoint, we believe that FAA's assumption that traffic demand and flight operations would not change as a result of system improvements poses a methodological limitation. FAA estimated that the airspace redesign will reduce airport delay by about 20 percent and annual operating costs by about $285 million once implemented (see Figure 7 for the calculation and context of the 20 percent delay reduction). Despite these estimated reductions, however, FAA assumed that traffic demand and flight operations would not change as result of system improvements.[66]

FAA's assumption that travel demand will not increase, however, is contrary to economic theory and FAA internal guidance.[67] Based on economic theory, because delays impose costs on airlines (by requiring them to operate their aircraft for longer periods and use more fuel) and on passengers (by requiring them to spend more time in delayed aircraft), a reduction in delays and an airline's operating costs would reduce the total price (or cost) of air travel.[68] The reduced total price of air travel would generally provide airlines with an incentive to lower fares (that is, they would be able to provide the same volume of service at a lower price) and because travelers will generally fly more at lower prices, the reduction in price could induce an increase in the amount of travel demanded by passengers, all else the same. FAA also failed to provide economic evidence to support its underlying rationale that traffic demand and flight operations will not change as a result of system improvements because the demand for air travel in the New York area is "price inelastic" (that is, relatively unresponsive to price changes).[69] FAA guidance, experts we interviewed, and our own previous work indicate that the responsiveness of travelers to changes in travel price depends on factors such as distance traveled, nature of the trip (nonbusiness versus business), and the availability of substitute travel modes (for example, rail). Specifically, FAA's benefit-cost guidance for capacity-related airport projects (for example, building a new runway) indicates that the demand for travel over shorter distances and travel by nonbusiness travelers is price elastic (that is, relatively responsive to changes in price).[70] In addition, an economist in the aviation industry whom we interviewed noted that air travelers who use the New York and Philadelphia airports have multiple travel options, especially options that compete with short-distance flights, and as a result, demand for air travel on these routes is relatively elastic. Our prior work also indicates that passenger rail competes with short-distance flights, such as the New York to Washington, D.C., market.[71]

We also question FAA's underlying rationale that flight operations would not increase as a result of the airspace redesign because the redesigned airspace does not increase overall system capacity. Specifically, FAA explained that the capacity of the entire airspace system—which includes the airspace, runways, and airport terminals—is determined by the element in the system with the least capacity, which, for the New York metropolitan area, is the capacity of the airport runways. Because this redesign does not increase runway capacity, FAA concluded that operations would not increase. However, FAA's National Airspace Capital Investment Plan states that an objective of this airspace redesign project is to increase capacity to meet projected demand and reduce congestion.[72] In addition, officials from the Port Authority of New York and New Jersey—which operates JFK, LaGuardia, and Newark—said that this redesign could increase capacity and that additional operations at these airports could be achieved. Finally, most of the experts that we interviewed believe that the capacity in the airspace would increase, which could allow for additional operations. Most of these experts, however, also stated that any increase in system capacity resulting from the redesign would likely be relatively small, such that any increase in new operations would also be small. One expert that we interviewed stated that there needs to be a major increase in the capacity, such as a new runway, before airlines would substantially increase their operations. Also, airlines could use a larger aircraft, with more available seats, instead of increasing the number of operations. While we believe this was a methodological limitation, it is uncertain what the net effect of the airspace redesign will be since the final results will depend upon the actions of the airlines and passengers in responding to the estimated delay reduction and operating cost savings.

FAA Did Not Fully Account for Future Use of New Technology in the Noise Analysis

FAA's assumptions in the noise analysis did not fully account for the future use of Area Navigation (RNAV) technology.[73] Specifically, FAA modeled RNAV procedures in the noise analysis of the mitigation measures for the preferred alternative, but did not model RNAV in the noise analysis used to compare the alternatives, which is inconsistent with the operational analysis. RNAV is designed to allow aircraft to fly a more precise track, thus examining the effect of RNAV on noise impacts could have provided more precise information for decision makers and the public.

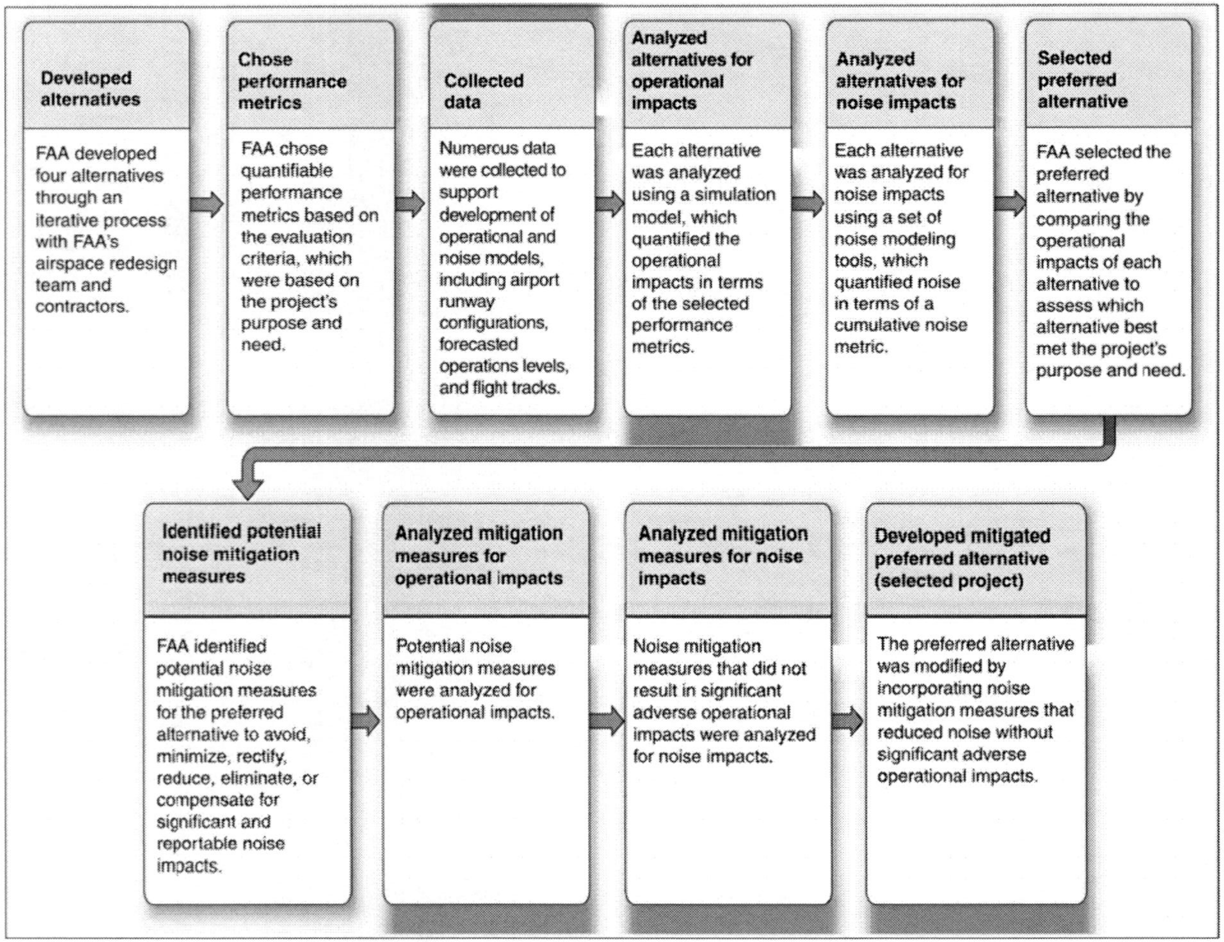

Figure 6. Simplified Representation of the Methodology Used by FAA to Assess Operational and Noise Impacts

Sources: GAO and FAA.

FAA projected that the selected project will reduce airport delays by 20 percent once fully implemented. While the scope of our study did not include a validation of FAA's estimates, we describe below additional information about FAA's projected delay reduction and how FAA calculated the 20 percent airport delay reduction for the selected project because the detailed calculations were not included in documentation that FAA provided to the public.

- The 20 percent delay reduction is the difference in airport delay in 2011 between the selected project and the Future No Action Alternative, not a reduction from today's airport delay.
- This airspace redesign addresses delays within FAA's control—that is, delays in the National Airspace System, as well as flexibility in severe weather. Thus, some causes of delay, such as air carrier and security delay, will not be reduced as a result of this project.
- The 20 percent airport delay reduction was calculated by combining the minutes of delay in clear weather (arrival delay, departure delay, and jet route delay) and severe weather (route flexibility). The difference in minutes of delay in 2011 between the selected project and the Future No Action Alternative was then translated into a percentage.
- Estimated delay reduction for the redesign is not equal for all airports—that is, delay reductions at some airports will be significantly greater than at others. In addition, some airports (most notably LaGuardia) will continue to experience significant airport delays. (See the table below for estimated airport arrival and departure delays.)

Airport Arrival and Departure Delays in 2011

Airport		Future No Action Alternative (minutes per flight)	Integrated Airspace with ICC (minutes per flight)	Difference (minutes per flight)	Percentage difference
JFK	Arrivals	3	1.7	-1.3	-43%
	Departures	11.7	9.4	-2.3	-20%
LaGuardia	Arrivals	53.7	49.4	-4.3	-8%
	Departures	74.7	65	-9.7	-13%
Teterboro	Arrivals	6.4	5.3	-1.1	-17%
	Departures	2.2	2.2	0	0%
Philadelphia	Arrivals	14.6	14.3	-0.3	-2%
	Departures	8.2	7.4	-0.8	-10%
Newark	Arrivals	34	26.7	-7.3	-21%
	Departures	20.1	13	-7.1	-35%

Figure 7. Description of FAA's Estimated 20 Percent Reduction in Airport Delay

Note: The table in fig. 7 does not include jet route delay or route flexibility used in FAA's calculation of the 20 percent airport delay reduction.

FAA said that it modeled RNAV for some mitigation measures of the preferred alternative because mitigation of noise impacts could be improved by using RNAV procedures. In the noise analysis used to compare the alternatives, however, FAA did not model RNAV because FAA said it did not know whether all aircraft using a particular route would be equipped with RNAV technology; FAA also did not predict which flight tracks airlines would choose. Thus, FAA officials said that they did not want to model anything that was not guaranteed to be implemented. Our review, however, showed that FAA had some indication as to which routes would be used with RNAV because some tracks in the operational models were designed to be RNAV compatible. An expert whom we interviewed further explained that because FAA had designed flight tracks to be more efficient for the purposes of this airspace redesign, it could be expected that airlines would use RNAV on those tracks that are most efficient. Furthermore, it would be reasonable to assume that RNAV will be used within the time frame of this airspace redesign because as one expert explained, RNAV reduces fuel burn and reduces flight time. FAA's own estimate reflects that about 80 percent of operations at the top 35 busiest airports in the National Airspace System are estimated to be RNAV capable.[74]

FAA officials also said that RNAV was not modeled in the noise analysis because the amount of dispersion depends on a route's designated accuracy value,[75] which is decided during development of the route procedure. According to FAA and its contractors, however, the effect of using RNAV in the noise analysis would have created a more concise concentration of flight tracks over specific geographic areas and would have concentrated noise under these tracks. Therefore, in most cases, RNAV may have resulted in fewer people impacted by noise but to a greater degree, depending on the location of the RNAV route. FAA included some RNAV procedures as noise mitigation measures, however. For example, when RNAV is used in the river approach to Philadelphia, the flight tracks are concentrated over the river, in effect reducing the number of people impacted by noise, as well as concentrating the higher noise levels over nonpopulated areas. Experts with whom we discussed FAA's use of RNAV said that examining the effect of RNAV procedures on the noise impacts could have provided more accurate information on the noise impacts. One expert, however, considered the FAA's assumption to not fully model RNAV in the noise analysis conservative given the uncertainty of the extent to which RNAV would become a key part of the airspace system.

FAA did not Assess the Effect of Noise Using Supplemental Noise Metrics

FAA only assessed noise impacts using a single cumulative noise metric that is required under FAA guidance, where additional metrics, not required by law or guidance, that measure noise in other ways—called supplemental noise metrics—would have provided information that may be more readily understood by decision makers and the public than the DNL metric.[76] As mentioned previously in this report, the Federal Interagency Committee on Aviation Noise recognized in 2002 that supplementing the DNL cumulative noise metric with alternative noise metrics would provide valuable information that is not easily captured by DNL, and the use of some supplemental noise metrics can provide a more meaningful estimate of physical impacts than a single DNL estimate. For example, Sound Exposure Level (SEL)[77] and Time-Above[78] are recognized as appropriate metrics for evaluating physical impacts associated with noise, such as sleep disturbance and speech interference.[79] According to FAA representatives, supplemental noise metrics were not calculated because the size and complexity of this regional airspace redesign would have made their inclusion in the EIS more confusing to the public. FAA has, however, calculated supplemental metrics, including SEL and Time-Above, in previous airspace redesigns.[80] While experts we interviewed recognized that presenting supplemental metrics in this EIS would have been challenging, they said that it was possible, and one expert explained that in some cases, it may have helped FAA more easily respond to public concerns. One expert said that limiting the information to discrete areas may have made presenting supplemental metrics in the EIS more manageable. For example, FAA could have reported single-event noise levels only for schools or the most common or loudest aircraft. In previous airspace redesigns where FAA calculated supplemental metrics, they reported the supplemental metrics only for areas that experienced noise impacts above a selected DNL value.

FAA did not Fully Assess Uncertainty Associated with Estimated Impacts of the Alternatives

FAA did not fully assess the uncertainty associated with the estimated impacts—that is, it did not consider a range of values for the key assumptions and inputs—even though some inputs are inherently uncertain.[81] Rather, when assessing the estimated impacts of the four alternatives, FAA generally used point estimates to represent traffic demand in 2006 and 2011, such as a point estimate for the future fleet mix and traffic levels. As a result, FAA's analysis implies greater precision about the redesign's impacts than is warranted. Assessing the uncertainty associated with the key assumptions and inputs would have provided a range of estimated impacts.

As FAA's own Handbook recognizes, questions are likely to be raised about analyses based on uncertain assumptions, and "one way to protect a study's integrity ... is to develop a few scenarios that reflect alternative assumptions for key uncertain elements,"[82] that is, perform a sensitivity analysis. We heard a range of opinions from experts we interviewed about assumptions made in the airspace redesign's demand forecast. Two experts said that the demand forecast did not fully account for new classes of aircraft, such as the change in use of air taxis or regional jets, while one expert noted that the forecast did not account for changes in the airlines' hub-and-spoke operations. Other experts disagreed with assumptions about airline yields or the effects of fuel costs. What most experts that we interviewed shared, however, was the belief that FAA should have assessed uncertainty by evaluating how changes in key assumptions made in the demand forecast would affect the estimated impacts of the alternatives.

FAA did, however, conduct a very limited sensitivity analysis to assess the uncertainty associated with one element of the forecasted fleet mix and one comparative analysis. According to FAA, based on the results of the limited sensitivity analysis, the relative ranking for the alternatives did not change. In 2006, FAA—in recognizing that the aviation industry had changed since the forecast was conducted—conducted a comparative analysis of some elements of its 2006 forecast with the observed levels in 2005 and some elements of the 2011 forecast of Philadelphia traffic to the 2012 forecast used for the Capacity Enhancement Plan.[83] This comparative analysis did not assess the uncertainty associated with estimated impacts, but, rather, assessed the accuracy of parts of the 2006 and 2011 forecasts developed for the airspace redesign. Through the comparative analysis, FAA determined that some of the forecasted elements were essentially correct (for example, the traffic counts for the 2006 forecast were 6 percent to 8 percent higher than the 2005 observed levels for the major airports in the region)[84] while some were inaccurate (for example, the estimated fleet mix).[85] Based on the results of this comparative analysis, FAA determined that the number of regional jets at Newark needed further analysis and as a result, conducted a very limited sensitivity analysis by increasing the estimated proportion of regional jets at Newark in 2006 and recalculating the operational impacts for each alternative in 2006. According to FAA, the results of the limited sensitivity analysis showed that the relative ranking of the alternatives was not affected by the change.

FAA did not Analyze Economic Impacts when Evaluating the Alternatives

The criteria used to evaluate the alternatives did not include economic impacts, such as implementation costs, the effect of the alternatives on the airlines, passengers' time value, or the effect of noise on the quality of life of residents living near the airports. According to experts interviewed, considering these impacts could have been helpful for decision makers because it would have provided more information for FAA to better assess whether the estimated benefits of the alternatives (relative to the Future No Action Alternative) justified the estimated costs. Experts we interviewed also said that a fuller discussion of the airspace redesign's potential costs may have increased stakeholders' understanding of FAA's decision and, in turn, made FAA's decision more palatable to the public. FAA, however, is neither required to, nor precluded from, considering these impacts in its decision under CEQ regulations.

Implementation costs

Some experts we interviewed noted that FAA did not assess implementation costs when comparing the alternatives with the Future No Action Alternative. According to FAA, because implementation costs, such as facility or training costs, would be the same for all alternatives, their inclusion in the operational analysis would not have affected FAA's decision. For example, FAA said that the cost for training air traffic controllers is an ongoing agency cost; that is, training will occur independently of the airspace redesign. However, because the selected project, unlike the other alternatives, will require a common automation platform,[86] its implementation costs[87] will be different from the other alternatives' implementation costs. Furthermore, although implementation costs were not explicitly analyzed when comparing the alternatives, FAA suggested in the airspace redesign's operational technical report that costs and benefits were assessed. Specifically, the operational technical report states that the Integrated Airspace with ICC is the only alternative "worth the effort and expense of implementing an airspace redesign of this magnitude."[88]

Other economic impacts

We also found that FAA did not assess other economic impacts, such as the effect of the alternatives on the airlines, passengers' time value, or the effect of noise on the quality of life of residents living near the airports. FAA representatives said they did not measure the economic impact on airlines or passenger time value because they were trying to keep the analysis "clean," and did not believe that those impacts were part of the NEPA process. In addition, FAA did not estimate the economic impacts of noise, in part because they are not required under CEQ regulations to conduct a benefit-cost analysis, and noise reduction was not part of the airspace redesign's purpose and need.[89] While methods to determine the economic costs associated with noise impacts are not firmly established, economists have used methods for assessing the economic impact of changes in noise.[90] For example, according to one expert we interviewed, FAA could have estimated property devaluation associated with noise using the DNL cumulative noise metric calculated in the noise analysis. Although not required under CEQ regulations, FAA would have had a more comprehensive picture of the trade-offs for each alternative had it assessed these economic impacts.

Uncertainty Analyses and Benefit-Cost Analyses Could Benefit Decision Makers and the Public

We identified two analyses that would have provided decision makers and the public with more comprehensive information. For example, an uncertainty analysis using a range of values for key assumptions and inputs, such as future demand and operations levels, would have provided decision makers and the public with a range of projected impacts. In addition, a benefit-cost analysis, which could have been used to assess the economic trade-offs between each airspace redesign alternative and the status quo, could provide more comprehensive information for decision makers to determine whether the estimated benefits associated with redesigning airspace justify the costs, and provide more transparency in the decision-making process to increase the public's understanding.

Uncertainty Analysis would Provide Information on the Range of the Redesign's Impacts

An uncertainty analysis that examined a range of key assumptions and inputs, such as the aviation demand forecast and the future use of RNAV, would have been useful. For example, a sensitivity analysis (a way to assess uncertainty) is used to change the values for key assumptions and to assess the effect on the estimated impacts, thereby providing decision makers and the public with information to understand the sensitivity of the results to changes in the assumptions. The analysis can be used to assess the uncertainty associated with the estimated impacts—that is, provide a range of projected impacts. The basic approach is to vary key assumptions, estimates, and forecasts systematically over appropriate ranges and to measure the impact on the results. In some cases, the relative impact of competing alternatives may be altered as a result of the sensitivity analysis, while in others it will not. As we stated in a prior section, FAA did not conduct a full sensitivity analysis on the demand forecast's key assumptions but only a limited analysis to assess uncertainty in the number of regional jets at Newark. FAA reported that the results of this limited sensitivity analysis showed that the relative ranking of the alternatives was not affected by the change.[91]

An uncertainty analysis is common for evaluations of projects that will have future impacts. For example, a 2007 report on the region's travel demand used a sensitivity analysis to assess the effect of low, baseline, and high scenarios for its 2025 forecast.[92] In addition, FAA's Handbook recommends a sensitivity analysis when conducting airspace studies and evaluating airspace alternatives. Specifically, the Handbook recommends a sensitivity analysis be performed "even if assumptions have a strong basis in precedence and factual information. Other assumptions are known to be inherently uncertain from the beginning because they are based on imperfect or incomplete information. This is clearly the case for all future events, such as the level and characteristics of the future traffic demand since the future cannot be predicted precisely."[93] Experts interviewed said that given the inherent uncertainty associated with projecting future events, when feasible, uncertainty should be assessed. Several experts told us that a sensitivity analysis could have been used for key factors in the operational analyses.[94] For example, a sensitivity analysis could have been performed on aircraft fleet operations at the region's smaller airports to account for the growing segment of business jet operations.

Benefit-Cost Analysis would Provide Information on Whether Estimated Benefits Outweigh Estimated Costs

A benefit-cost analysis, which can be used to assess the benefits and costs associated with each alternative compared to the status quo, would have provided decision makers and the public with more comprehensive information on whether the estimated benefits of the selected project justified its costs. Benefit-cost analyses are used to identify the alternative, if any, that maximizes the net benefits (benefits minus costs) relative to the status quo. The alternative that maximizes net benefits is economically justified. The analysis also can be used to identify how the benefits and costs associated with each alternative are distributed across different subpopulations or income groups.

FAA stated that it did not conduct a benefit-cost analysis for the airspace redesign because the agency was not required to do so.[95] GAO believes, based on prior work, that benefit-cost analyses can provide valuable information for decision makers and should be considered in analyzing transportation investments.[96] While conducting benefit-cost analyses requires an investment in time and resources, FAA uses benefit-cost analyses in making other investment decisions, such as the Next Generation Air Transportation System (NextGen) projects. Moreover, FAA has guidance for conducting benefit-cost analyses for airport projects and investment analyses for air traffic control procurements.[97] FAA requires benefit-cost analyses in granting discretionary airport improvement program funds and for investment analysis for air traffic control procurements. In addition, the Office of Management and Budget requires benefit-cost analyses for new regulations. The Air Traffic Organization (ATO), which includes the division that develops airspace redesigns, also includes a division that is charged with assessing investments for the air traffic control system. FAA's Office of Aviation Policy and Plans conducts benefit-cost analyses on investment and regulatory decisions and discretionary Airport Improvement Program grants.[98]

Furthermore, only after FAA selected the final alternative did it assess the economic benefits associated with the selected project. Specifically, in describing the benefits of the airspace redesign to the public, FAA stated that the selected project would result in 12 million minute delay savings (see Figure. 8 for further discussion) and a reduction of up to $285 million[99] in annual operating costs, when compared to the Future No Action Alternative. These benefits were calculated using FAA's economic values for investment decisions, which are used in FAA's benefit-cost analyses for investment and regulatory decisions.[100] However, the dollar amounts for delay savings were done only for airline operating costs and not for others such as delay savings to passengers. Moreover, because FAA did not estimate costs, the information on benefits cannot be used to assess whether the selected project is economically justified.

FAA's limited economic analysis focused only on select benefits of only one alternative—the selected project—and does not account for costs. A comprehensive benefit-cost analysis would have increased transparency about the range of potential economic impacts of the airspace redesign. For example, a benefit-cost analysis that estimated the economic impact on passengers, airlines, and residents living near the airports, would have allowed a decision maker to assess which alternative, if any, would maximize net benefits, and to identify how the benefits and costs are distributed among population or income groups. As one expert explained, it is difficult to identify from the operational analysis the beneficiaries of the system improvements. For example, system improvements, such as reducing delay, benefit passengers and airlines, but other improvements, such as reduce voice communications and balance controller workload, benefit the ATC system.

Experts we spoke with indicated that a benefit-cost analysis for the airspace redesign would have been useful. For example, one expert noted that the list of system improvements and metrics in the alternatives analysis focused on impacts related to efficiency, while the implementation costs were not included. In addition, if the economic impacts of noise had been included, costs could have varied by alternative. For example, FAA estimated that the Ocean Routing Airspace Alternative would result in fewer people impacted by noise compared to other alternatives. If the economic effect of noise, approximated by a change in home values due to aviation noise, had been taken into account, then an alternative like the Ocean Routing Airspace Alternative might have performed better than the other alternatives on this metric. In addition, the potential economic effect on passengers and airlines also would have been assessed in a comprehensive benefit-cost analysis. For example, the effect of delay on passengers could have been assessed using the value of passenger time lost or gained as a result of the airspace redesign.

LACK OF DETAILED IMPLEMENTATION PLAN RAISES QUESTIONS ABOUT TIME FRAMES AND COSTS

Based on the information available to date, we are unable to determine whether FAA will meet its projected 5-year implementation time frame, what the final project configuration and implementation costs will be, and any details regarding FAA's postimplementation evaluation plan. For example, FAA has not yet developed a detailed implementation plan that includes a schedule of when the new redesigned airspace will be in place. Additionally, FAA has not yet determined the type or amount of equipment or software that will be needed to implement the project.

FAA stated in press releases that there will be a projected annual 12 million minutes in delay savings. We present this description because FAA did not include an explanation of the calculation in the documents it released to the public.

- This measure was developed as a means to communicate the delay benefits of the redesigned airspace to the public. FAA officials said that because this measure only reflects delay, it is not a useful measure in evaluating the different alternatives. Furthermore, it was developed after FAA made its decision.

- The 12 million minutes is not a straight multiplier of the average minutes delay savings—which FAA estimates to be 3 minutes of arrival delay per flight—by the number of operations at the airports. Rather, the 12 million minutes was calculated from 4 performance metrics—traffic-weighted arrival delay, traffic-weighted departure delay, departure airspace delay, and delay savings from flexibility-of the 13 metrics used in the operational analysis to compare the alternatives. FAA multiplied the daily per flight savings by the number of affected flights and then annualized the savings for each of the four metrics. The total savings calculated for all four metrics was approximately 12 million minutes (see table below).

- In one press release, FAA stated that the 12 million minutes savings were for the four major metropolitan airports—JFK, LaGuardia, Newark, and Philadelphia. However, the operational analysis included other satellite airports.

Calculation of 12 Million Minutes Savings

Metric	Minutes saved	Number of operations	Annualizing factor	Annual savings (minutes/year)
Minutes saved per flight for arrival	2.948	3,558	368	3,860,145
Minutes saved per flight for departures	4.113	3,566	368	5,398,744
Minutes saved per flight for departure airspace delay	0.49	3,566	368	643,021
Minutes saved per flight for routing flexibility	12.60	3,566	36	1,617,538
Total minutes saved				**11,519,448**

Figure 8. Description of the 12 Million Minutes Saving in Delay Savings

Sources: GAO; FAA (table).

Lack of Detailed Implementation Plan Raises Questions on FAA Meeting Its 5-Year Completion Goal

FAA is in the process of preparing an implementation plan, and therefore, we are unable to determine whether FAA can meet its projected 5-year implementation time frame, overcome potential obstacles to implementation, and transition from the current airspace structure to the redesigned structure. FAA is responsible for ensuring that the selected project is implemented in a manner that avoids any interruption in air traffic services and minimizes disruptions to air traffic. To ensure that new airspace redesign projects are successfully implemented, FAA's Handbook lays out specific activities for transitioning between the current airspace structure and the redesigned airspace. Based on our prior work[101] and the experts we interviewed, we have identified three specific activities recommended in FAA's Handbook and one additional activity emphasized by our experts that we believe are important to include in an implementation plan to successfully implement the redesigned airspace.

- *Scheduling. FAA's Handbook recommends preparing a schedule that identifies the exact time that the new redesigned airspace will be in place so that all affected FAA regions and facilities, airports, airlines, air traffic controllers, and general aviation pilots can prepare for the change. For example, a major component of the redesigned airspace is the integration of the airspace, which will allow air traffic controllers to use the terminal separation rules over a larger geographical area. Although FAA acknowledges that integrating the airspace will require new procedures and potentially reallocation of staff, there is no detailed information on when these procedures will be implemented, how ATC staff may be reallocated, and what type of air traffic controller training will be required. Without this information, it is difficult for the agency to clearly track and report progress, as well as to ensure timely implementation. In our prior work,[102] we have highlighted the importance of developing an implementation timeline to build momentum and show progress from day one. For example, to help federal agencies implement successful transformation of their cultures in a transparent and accountable manner, a key implementation step is to make implementation goals and timelines public.*

- *Risk mitigation. According to the Handbook, as part of its implementation plan, FAA should identify potential risks and develop a risk mitigation plan to minimize the impact if the risk does occur. FAA, however, has not developed risk mitigation strategies to deal with potential challenges that may arise during the redesign's implementation. Without details on proposed risk mitigation, it is unclear whether FAA has taken steps to identify possible risk factors. FAA officials noted that there are a number of separate FAA projects and initiatives that, while not directly linked to the airspace redesign project, will influence how FAA implements the airspace redesign. For example, FAA has not decided whether it will house the new common automation platform in existing facilities, a new facility, or a consolidated facility. In some of our prior work,[103] we have recommended that agencies should develop a risk analysis to show the potential impacts if task and milestones are not achieved, develop methods to accurately evaluate and measure the progress of implementation, and develop contingency plans should task and milestones not be met.*

FAA does not have such plans in place for implementing the redesign. Furthermore, if there is no consideration of risk mitigation, FAA may be unable to minimize potential risks.

- *Transition planning. The Handbook also recommends that FAA identify the steps that need to be taken to transition between the current airspace to the redesigned airspace in a transition plan that includes key decision points along the way. Transitional plans provide details on the activities and time schedules for the transition from the current airspace to the redesigned airspace. The purpose of these plans includes ensuring that services associated with the airspace and airports continue throughout implementation and that the multiple users of the airspace—such as pilots, FAA personnel, and air traffic controllers—are aware of the ongoing changes. Given this redesign's high level of complexity and multiple implementation stages, transitional planning could be beneficial to FAA's implementation. However, FAA has not developed a transitional plan for the redesign's implementation. We have highlighted in a prior GAO report[104] that FAA should include proper transition planning when implementing highly technical and complex projects that involve numerous stakeholders, such as NextGen. For example, in February 2008, we testified that FAA faces a number of management challenges—such as better aligning costs with revenue—in its transition to NextGen.[105]*

- *Monitoring. It is important for FAA to monitor the implementation throughout the transition process to ensure that FAA is actually implementing the changes in the airspace that were identified in the selected project. We have reported in prior GAO work[106] that monitoring efforts can help agencies transition to new systems. For example, as FAA transitions to NextGen, we believe it will be critical for FAA to monitor and address equipment outages to ensure the safety and efficiency of the current ATC system. In addition, experts we spoke with also believe that monitoring is an important step for implementing the airspace. For example, one expert interviewed by GAO stated that FAA should get a current baseline of information on the airspace, and continue to monitor and evaluate throughout implementation. Another expert interviewed by GAO also stated that both operational and noise impacts should be assessed (as part of monitoring and evaluation) to further enhance management of the implementation. Although FAA has already begun implementing the redesign, it is unclear how the agency will monitor and evaluate the process since there is no monitoring plan in place. Periodic monitoring and evaluation will both ensure that FAA is complying with the operational and environmental objectives outlined in the selected project, as well as increase the project's accountability and transparency.*

Final Project Configuration and Costs Are Unknown

FAA has not determined the final project configuration and costs. Based on our prior work,[107] it is critical for an agency to have a full understanding of project costs to ensure that implementation is successful. For example, by understanding project costs, agencies can ensure they have the required resources. However, FAA has not yet determined the costs due, in part, to two reasons. First, although FAA officials have stated that they will be using currently available technology and software, FAA has not determined the type or amount of equipment and software that will be needed to implement the common automation platform for the ICC. Second, it is still unclear whether the ICC will be housed in existing FAA facilities, a new facility, or in a consolidated facility. Under FAA's Facility Replacement Program, FAA is reviewing whether to build a new TRACON facility for the New York region. However, FAA officials estimated that the final decision as to whether to approve the design and construction of a new TRACON will be made sometime in September 2008, but it will not include costs estimates. Cost estimates will require FAA to make specific decisions on the location of the TRACON, the size of the facility, and any supporting infrastructure requirements. If FAA approves the construction of a new TRACON, the final cost estimate for this new facility will not be available until September 2011 when the contract is to be awarded, and the new facility is not projected to be completed until September 2015—3 years after FAA has estimated full implementation of the selected project.

An Adaptive Management Strategy May Help FAA Successfully Evaluate the Implementation of the Redesign

Although the redesign's implementation is planned to take multiple years, FAA's Handbook offers few actions and steps for the agency to use for a postimplementation evaluation of the redesign. Given that the redesign represents a complex and comprehensive change to the region's airspace, it is important for FAA to determine how it plans to conduct evaluations of the redesign. One potential strategy is adaptive management. As defined by the National Research Council, adaptive management is a process that promotes flexible decision making in the face of uncertainties, as outcomes from management actions and other events become understood. Adaptive management is recognized by CEQ and other federal agencies as a valuable strategy.[108] For example, in 2007, CEQ issued a guide[109] that recognizes the value of adaptive management when there are uncertainties in the prediction of the impacts or outcome of project implementation. GAO also has recommended adaptive management, and practices like it, as a strategy for other federal agencies. For example, we recommended that the U.S. Department of Agriculture adopt a systematic application of adaptive management principles and develop a monitoring program for Yellowstone's bison[110] to improve its overall management of the program. Adaptive management elements that FAA could use include implementing a monitoring program to systematically obtain information about the management objectives identified in the selected project and communicating and engaging the results with key stakeholders. This approach also would emphasize FAA's accountability and transparency to its stakeholders. In addition, a 2004 FAA paper describes elements for using adaptive management.[111] One expert interviewed by GAO stated that a benefit of using an adaptive management strategy as a review tool, both to assess operational and environmental impacts, is the effect it could have on the design and development of future airspace redesigns.

CONCLUSIONS

When conducting its airspace redesign, FAA complied with legal requirements under NEPA and environmental justice directives for the identified five key issues. In addition, FAA's methodology to assess operational and noise impacts was reasonable. Nonetheless, we found that FAA's methodology contained some limitations that if addressed in future efforts, could improve FAA's decision making and the public's understanding of FAA's decision and process. In particular, FAA could undertake two types of analyses—uncertainty analyses and benefit-cost analyses—in future airspace redesigns to help avoid limitations similar to those identified during our evaluation. An uncertainty analyses would provide decision makers and the public information on the extent to which estimated impacts may vary under alternative assumptions. As highlighted in our study, the realization of the redesign's operational benefits are dependent on the accuracy of key assumptions—particularly the aviation demand forecast. FAA has already established through a comparative analysis that the traffic counts for the 2006 forecast are higher than the 2005 observed levels; thus, while an uncertainty analysis might not have resulted in a change in the relative ranking of the alternatives, our confidence that the level of estimated benefits will be realized is reduced. A benefit-cost analysis, although not required, could provide more comprehensive information for FAA to assess whether the estimated benefits associated with redesigning airspace justify the costs. More information on the trade-offs would also increase transparency in FAA's decision for the public, thus, in turn, likely making the decision more understandable. Although a benefit-cost analysis would require an investment in time and resources, FAA conducts these types of analyses for other investment decisions, such as ATC procurements and new regulations, and has specific guidance for them. In addition, economists have used methods to estimate the economic impact of changes in noise, and supplemental noise metrics, which provide different information about noise impacts not captured by DNL, are recognized as appropriate methods for analyzing physical impacts.

Our study also raises concerns about FAA's future actions for this airspace redesign. Despite implementing the initial steps of the redesigned airspace last December, FAA has not published a detailed implementation plan. The lack of such a plan raises concerns that FAA may not meet its estimated implementation time frame, be able to deal with potential obstacles, or be able to transition from the current airspace structure to the redesigned structure. Furthermore, it is important for FAA to determine how it plans to evaluate the implementation, which would help FAA ensure that the operational and environmental objectives are met and increase the project's transparency and accountability. One strategy used by other federal agencies for evaluating a project's implementation is the adaptive management strategy. Adaptive management is a strategy that recognizes the need for flexibility and would allow for modification of the redesigned airspace if the results of future research indicate a need for change, as long as these changes do not significantly affect the quality of the human environment. This strategy would include implementing a monitoring program to collect information and communicating and engaging with key stakeholders.

RECOMMENDATIONS FOR EXECUTIVE ACTION

To improve FAA's effectiveness and accountability in implementing the New York/New Jersey/Philadelphia Airspace Redesign, we recommend that the Secretary of the Department of Transportation direct the Acting Administrator of the Federal Aviation Administration to take the following two actions for the New York/New Jersey/Philadelphia Airspace Redesign:

- *Develop and follow a detailed implementation plan that includes a time and cost schedule, risk mitigation plan, transition planning, and monitoring and evaluation plan.*
- *Follow a postimplementation evaluation plan that includes an adaptive management strategy for monitoring implementation of the redesign and communicating the results to key stakeholders.*

To improve FAA's effectiveness in conducting future airspace redesigns, we recommend that the Secretary of the Department of Transportation direct the Acting Administrator of the Federal Aviation Administration to take the following actions in developing and implementing future airspace redesigns:

- *Conduct an uncertainty analysis of key assumptions and inputs—particularly on elements within aviation demand forecasts. The analysis should be used to assess the extent to which the estimated impacts for the airspace redesign alternatives would change using different values for key assumptions and inputs and to provide information on the level of confidence in the project's estimated impacts and the relative ranking of the alternatives.*
- *Conduct a benefit-cost analysis for the purpose of assessing the economic effect of alternatives for airspace redesigns (including the status quo). Such an analysis should include an assessment of the key impacts associated with redesigning the airspace, including implementation costs and, as appropriate, the economic effect associated with noise.*

Susan A. Fleming
Director, Physical Infrastructure Issues

Susan D. Sawtelle
Managing Associate General Counsel

APPENDIX I: OBJECTIVES, SCOPE, AND METHODOLOGY

In assessing the Federal Aviation Administration's (FAA) airspace redesign of the New York/New Jersey/Philadelphia region, the report addresses the following questions: (1) To what extent did FAA follow applicable key legal procedures and requirements in conducting its environmental review? (2) To what extent was the methodology used by FAA to assess the operational and noise impacts reasonable? (3) What is the likelihood that FAA will meet its project time frames and costs of implementing its airspace redesign project?

To address our three research questions, we reviewed documents associated with the New York/New Jersey/Philadelphia Airspace Redesign, including the draft and final environmental impact statements (EIS) and selected appendices, underlying technical reports and analyses, and the Administrative Record filed by FAA with the court in the pending litigation. We also obtained and analyzed information from a variety of other sources, including federal laws, regulations, executive directives, and court decisions; FAA and the Department of Transportation (DOT) Orders and FAA and Council on Environmental Quality (CEQ) guidance; the final EISs for three previous FAA-sponsored airspace redesigns determined to be similar to this project;[112] prior GAO work; and other aviation evaluations and studies. We interviewed officials from DOT, the Environmental Protection Agency, the Port Authority of New York and New Jersey, City of Philadelphia Department of Aviation, and FAA's headquarters and eastern regional offices, including representatives of FAA's airspace redesign team and its principal contractors for the operational and environmental analyses. We also contacted other stakeholders, including a trade organization for the aviation industry, the national air traffic controller's union, and local governments and community groups in the region.

To assess the extent to which FAA followed key legal procedures and requirements, we reviewed applicable federal laws, regulations, executive directives, and court decisions, as well as FAA and DOT Orders and FAA and CEQ guidance, to establish criteria by which to compare FAA's actions. We then conducted a legal analysis to determine whether FAA complied with five key issues for the airspace redesign to include the statement of the project's purpose and need, evaluation of alternatives, consideration of the project's environmental effects, public involvement, and environmental justice matters. We selected these five issues based on public concerns raised during and after the EIS process, congressional interest, the views of experts we interviewed, and GAO's evaluation of the range of concerns presented. To examine the extent to which the methodology used by FAA to assess the operational and noise impacts was reasonable, we reviewed FAA's methodology for conducting the operational and noise analyses. We compared FAA's methodology to criteria we established through our review of federal policy, FAA guidance, prior GAO reports, and standards from the aviation and analytical community, including generally accepted economic principles and practices, and statistical principles.

In addition, with the assistance of the National Academy of Sciences, we identified 11 experts in the fields of EIS policies and procedures, airspace operations and system modeling, and aircraft noise measurement and mitigation. These experts reviewed selected portions of the EIS related to FAA's operational and noise analysis. We interviewed these experts within their area of expertise to obtain their views on the extent to which the FAA followed applicable procedures and requirements, and on the methodology used by FAA to assess the operational and noise impacts. (Table 2 identifies the list of participating experts.)

To determine the likelihood that FAA will meet its project time frames and costs of FAA's airspace redesign, we reviewed documents associated with the redesign's impacts, time frames, and direct costs, and met with FAA to discuss plans for project implementation. We also reviewed prior GAO work related to environmental reviews and project implementation.

Table 2. List Of Experts Providing Input During Our Review

Expert	Area(s) of expertise
Hamsa Balakrishnan, Assistant Professor of Aeronautics and Astronautics, Massachusetts Institute of Technology	Airspace operations and system modeling
Mike Ball, Orkand Corporation Professor of Management Science, Decision and Information Technologies Department, University of Maryland	Airspace operations and system modeling
William Dunlay, Director, Jacobs Consultancy	EIS policies and procedures
Charles Etter, Staff Scientist, Gulfstream Aerospace Company	Aircraft noise measurement and mitigation
Bill Jeffers, Senior Director, ARINC, Inc. (retired)	Airspace operations and system modeling
Margaret Jenny, Chief Executive Officer, MJF Strategies, LLC	Airspace operations and system modeling
Vincent Mestre, Principal, Mestre Greve Associates, Inc.	Aircraft noise measurement and mitigation
Clint Oster, Professor, School of Public and Environmental Affairs, Indiana University	Airspace operations and system modeling
Neil Planzer, Vice President of Air Traffic Management Strategy, The Boeing Company	Airspace operations and system modeling
John Putnam, Partner, Kaplan Kirsch & Rockwell	EIS policies and procedures
Ian Waitz, Jerome C. Hunsaker Professor of Aeronautics and Astronautics, Massachusetts Institute of Technology	Aircraft noise measurement and mitigation

Source: GAO.

As agreed with you, we did not conduct a new EIS, or develop and analyze new alternatives to the airspace redesign. We conducted this performance audit from July 2007 to July 2008 in accordance with generally accepted government auditing standards. Those standards require that we plan and perform the audit to obtain sufficient, appropriate evidence to provide a reasonable basis for our findings and conclusions based on our audit objectives. We believe that the evidence obtained provides a reasonable basis for our findings and conclusions based on our audit objectives.

APPENDIX II: FAA'S LEGAL COMPLIANCE WITH KEY NEPA REQUIREMENTS AND ENVIRONMENTAL JUSTICE DIRECTIVES

Introduction and Summary of Conclusions

As part of GAO's review of the FAA's airspace redesign for the New York/New Jersey/Philadelphia region, we examined FAA's legal compliance with certain requirements of the National Environmental Policy Act (NEPA), 42 U.S.C. §§ 4321 *et seq.*, and with the environmental justice directives of Executive Order 12898, "Federal Actions to Address Environmental Justice in Minority Populations and Low-Income Populations," 59 Fed. Reg. 7629 (Feb. 16, 1994). Specifically, based on public concerns raised during and after the EIS process,[113] congressional interest, the views of experts we interviewed, and GAO's evaluation of the range of concerns presented, we reviewed FAA's compliance with respect to five issues:

- *whether FAA's purpose and need statement in its EIS for the airspace redesign was reasonable;*
- *whether FAA developed a reasonable range of alternatives to the redesign and rigorously explored those alternatives in the EIS;*
- *whether FAA reasonably decided not to analyze environmental effects of a potential increase in growth resulting from the redesign;*
- *whether FAA reasonably involved the public in its environmental review process; and*
- *whether FAA appropriately considered environmental justice issues in its environmental review process.*

Based on the information available to us and for the reasons discussed below, we conclude that FAA met NEPA and Executive Order 12898 requirements in conducting the airspace redesign with respect to these issues.[114]

BACKGROUND

As discussed in greater detail in this report, in 1998, FAA launched a comprehensive national airspace redesign effort to more efficiently and safely move aircraft through the nation's airspace system. As a part of this program, FAA initiated the New York/New Jersey/Philadelphia airspace redesign in 1999. For a federal project of this magnitude, FAA must comply with NEPA and implementing regulations issued by CEQ, as well as FAA's own NEPA Orders and guidance.[115] FAA also must follow environmental justice directives set forth in Executive Order 12898. After determining that NEPA required an EIS for the project, FAA issued a draft EIS in December 2005. Following public hearings and a public comment period, FAA issued a final EIS in July 2007 and, based on the final EIS, issued a Record of Decision in September 2007, designating one of its redesign alternatives as the selected project. FAA has begun implementing this project and anticipates that it will take 5 years to complete implementation. Among other things, an EIS must: (1) address the purpose of and need for the proposed action;[116] (2) identify alternatives to the proposed action;[117] and

(3) present the reasonably foreseeable environmental impacts of the proposed action, including the direct and indirect effects and cumulative impacts.[118] Furthermore, throughout the environmental review process, an agency must make diligent efforts to involve the public[119] and, to the greatest extent practicable and as appropriate, must identify and address any disproportionately high and adverse human health or environmental effects of its activities on minority and low-income populations.[120] We address FAA's compliance with each of these requirements in turn.

Issues and Analysis

1. Was FAA's purpose and need statement reasonable?

As discussed below, FAA's statement of the purpose and need for the airspace redesign was reasonable and therefore complied with NEPA requirements. It satisfied the statutory and regulatory requirements as they have been applied by the courts—the statement was not too narrowly or broadly defined—and it reasonably excluded noise reduction as a purpose.

NEPA Requirements

CEQ regulations implementing NEPA require an agency to include in the EIS a statement specifying the purpose of and need for the project—commonly referred to as a "purpose and need statement." In particular, 40 C.F.R. § 1502.13 requires that the agency include a statement to "briefly specify the underlying purpose and need to which the agency is responding in proposing the alternatives including the proposed action." Further, FAA Order 1050.1E, section 506(d) requires the purpose and need statement to "present[] the problem being addressed . . . and essentially provide[] the parameters for defining a reasonable range of alternatives to be considered." As the courts have explained, an agency "bears the responsibility for defining at the outset the objectives of an action." *Citizens Against Burlington, Inc. v. Busey*, 938 F.2d 190, 195-96 (D.C. Cir. 1991), *cert. denied*, 502 U.S. 994 (1991). The scope of the project, as articulated by the agency, informs the alternatives it considers and guides its decision-making process.

In determining whether an agency has complied with NEPA, courts grant substantial deference to an agency's actions as required by the Administrative Procedure Act.[121] With respect to the purpose and need statement in particular, courts generally apply a "rule of reason" and uphold the statement as long as it is reasonable. *City of Alexandria v. Slater*, 198 F.3d 862, 867 (D.C. Cir. 1999), *cert. denied*, 531 U.S. 820 (2000). The agency can neither define a project's purpose and need so narrowly that there is only one logical alternative to accomplish it, nor so broadly that an infinite number of alternatives would accomplish its goals. *Citizens Against Burlington*, 938 F.2d at 196 (upholding FAA approval of airport expansion). An agency complies with NEPA if it meets these procedural requirements.

FAA Actions

FAA initiated a redesign of the New York/New Jersey/Philadelphia region's airspace as a first step in its national airspace redesign effort to accommodate air traffic growth while maintaining safety and mitigating delays, and to accommodate changes in the types of aircraft using the airspace system. FAA's basic objective was to create a system to better manage traffic and capacity in the region, and to replace the air traffic system designed in the 1960s. FAA noted that while the volume of traffic and the type of aircraft used have changed considerably since that time, the basic structure of the airspace has stayed the same, and emerging technologies have not been taken into account.

To increase the efficiency and reliability of the airspace structure and air traffic control (ATC) system in the New York/New Jersey/Philadelphia region, FAA began by identifying a formal purpose and need for the airspace redesign. To do this, MITRE's Center for Advanced Aviation System Development (CAASD)—a Federally Funded Research and Development Center sponsored by FAA—identified problems and performance shortfalls. In a parallel effort, FAA developed focused leadership teams, composed of experts from throughout the United States, who began a systematic effort to identify airspace inefficiencies and solutions to ensure national integration of local efforts.[122] Based on these efforts, FAA developed the following purpose and need statement:

> "The purpose of the airspace redesign is to increase the efficiency and reliability of the airspace structure and ATC system. The need is to accommodate growth in aircraft operations while maintaining safety, mitigating delays, and accommodating changes in the types of aircraft using the System."[123]

FAA has identified these same objectives—increased efficiency and safety—for several of its previous airspace redesign projects, such as the Potomac Consolidated Terminal Radar Approach Control (TRACON)[124] in the Washington/Baltimore area and the Chicago Terminal Airspace Project.

FAA's purpose and need statement for the New York/New Jersey/Philadelphia airspace redesign established the overall criteria against which different alternatives were examined (see section 2 below). Notably, FAA did not include noise reduction as part of the project's formal purpose and need. As FAA explained in the final EIS: "Noise reduction is not a component of the purpose and need for the proposed action. Reduction of noise is not appropriately identified as a purpose because it is not FAA policy to reroute aircraft to reduce noise levels in one community at the expense of another."[125] Rather, according to FAA, the "first consideration and highest priority in defining the Purpose and Need for any proposed action is to serve the public interest by exercising its authority to assign, maintain, and enhance safety and security of the national airspace (per 49 U.S.C. § 40101(d))."[126] This is consistent with FAA's statement of its mission: to provide the safest, most efficient aerospace system in the world.[127]

Nonetheless, there was considerable public controversy regarding the potential noise effects of FAA's redesign and, as the project progressed, congressional committees urged FAA to consider noise impacts and mitigation where appropriate.[128] FAA took several such actions, beginning with a commitment in its early scoping meetings to reduce noise where possible—for instance, by increasing altitudes, dispersing or concentrating routes where appropriate, reducing fly time, and routing aircraft over less noise-sensitive areas.[129] Additionally, FAA included a detailed analysis of noise impacts for each alternative considered in detail[130] and included one alternative—the Ocean Routing Alternative—that would have reduced noise overall. Finally, FAA included proposed mitigation measures along with its preferred alternative, and issued a separate 62-page noise mitigation report.[131] As FAA noted in the final EIS, however, the airspace redesign was not a cure-all for aircraft-related noise problems in the region.[132]

Analysis

FAA's purpose and need statement for the airspace redesign complied with NEPA requirements. FAA's statement defined the parameters of the project, as required by 40 C.F.R. §1502.13. It also satisfied the requirements courts have developed in evaluating agency statements of purpose and need. FAA did not define the project's purpose so broadly that it generated an infinite number of alternatives, as demonstrated by its exclusion of such options as airport infrastructure modification, congestion management measures, and alternate modes of transportation[133] (see section 2 below). Nor did FAA define the purpose so narrowly that only one alternative would be satisfactory, as evidenced by the fact that it could be met by multiple alternatives (see section 2). While FAA ultimately selected a single alternative that best met the project's purpose and need, more than one alternative would have done so.

FAA also acted reasonably in excluding noise reduction as part of the project's formal purpose and need statement. As noted, in analyzing the environmental impacts of the proposed redesign, FAA gave substantial consideration to possible noise effects. It was nevertheless reasonable to exclude noise reduction as a formal objective of the redesign, because NEPA "[does] not require agencies to elevate environmental concerns over other appropriate considerations." *Baltimore Gas & Electric Co. v. NRDC*, 462 U.S. 87, 97 (1983); *see also Strycker's Bay*, above, 444 U.S. at 227. Rather, through NEPA's requirement that agencies prepare an EIS for all major federal actions significantly affecting the environment, the statute requires only that agencies consider the environmental impacts of their actions as part of their decision making, not that they use environmental impacts as the deciding factor. *Kleppe v. Sierra Club*, above, 427 U.S. at 409.

Courts have approved other purpose and need statements that excluded environmental concerns. In *City of Alexandria v. Slater*, above, for example, the D.C. Circuit evaluated the purpose and need statement prepared by the Federal Highway Administration (FHWA) for the planned replacement of the Woodrow Wilson Bridge in the Washington, D.C., area. The agency's statement focused primarily on traffic and safety concerns, rather than on environmental concerns such as noise. In finding the agency's purpose and need statement to be reasonable, the court stated that "[t]he proper question to ask at the outset of a NEPA inquiry is not whether the [FHWA] focused on environmental goals but rather . . . whether its stated objectives were reasonable. It seems rather obvious to us that it is not unreasonable in articulating its objectives for an agency to 'focus primarily on transportation and safety issues' when replacing a massively congested and structurally unsound bridge." 198 F.3d at 867-68. Similarly, in *Citizens Against Burlington*, above, the D.C. Circuit approved FAA's EIS for a major airport expansion that would cause significant noise impacts, even though

FAA's stated purpose and need focused only on creating a new cargo hub, and did not mention noise mitigation.

By contrast, in *California v. DOT*, 260 F. Supp. 2d 969 (N.D. Cal. 2003), the court rejected FAA's purpose and need statement in its environmental assessment of a major airport expansion, because it was so narrow that it effectively eliminated consideration of any significant environmental effects. While FAA stated that the purpose of the expansion was simply to "provide the necessary runway length to safely allow [larger aircraft] . . . to operate at the Airport," and the court agreed that such a narrow purpose might result in no significant environmental impacts, in fact, the record showed the project's very purpose was to induce growth by attracting thousands of additional tourists to a local ski resort. The court therefore rejected FAA's "myopic view of the airport project" and, based on this and other reasons, required a full EIS to be prepared. *Id.* at 974.

As in *City of Alexandria* and *Citizens Against Burlington*, the issue with respect to the New York/New Jersey/Philadelphia Airspace Redesign is not whether FAA included environmental concerns in its purpose and need statement, but whether the agency's purpose and need statement was reasonable. As discussed above, we conclude it was, because it identified the problem being addressed and thus essentially provided the parameters for defining a reasonable range of alternatives to be considered, as the regulations and FAA's Orders require.[134] This is not a case like *California v. DOT*, where the facts belie the agency's stated purpose.

Finally, the congressional committee reports directing FAA to consider or mitigate the project's noise impacts[135] did not create a legal requirement to include noise reduction or mitigation as part of the project's formal purpose and need. The D.C. Circuit noted in *Citizens Against Burlington*, above, that in formulating a statement of purpose and need, agencies "should always consider the views of Congress, expressed, to the extent that the agency can determine them, in the agency's statutory authorization to act, as well as in other congressional directives." 938 F.2d at 196. FAA did consider the "views of Congress" in developing the project's purpose and need—it complied with the mandate in 49 U.S.C. § 40101(d) to give the highest priority to enhancing the safety and efficiency of the national airspace. FAA also considered the congressional report statement directing FAA to address noise issues it its New York/New Jersey/Philadelphia Airspace Redesign, by conducting detailed technical noise analyses and including noise mitigation measures for the selected project. These report directives did not, however, legally bind FAA to include noise mitigation as a purpose of the redesign because they were not contained, or incorporated by reference, in enacted legislation. *See generally Cherokee Nation of Oklahoma v. Leavitt*, 543 U.S. 631, 646 (2005); *Lincoln v. Vigil*, 508 U.S. 182, 192 (1993); GAO, *Consolidated Appropriations Act, 2008—Incorporation by Reference*, B-316010 (Feb. 25, 2008).

4. Did FAA develop a reasonable range of alternatives to the proposed redesign and rigorously explore those alternatives in the EIS?

As discussed below, FAA developed a reasonable range of alternatives and rigorously explored those alternatives, as required by NEPA. The agency included a no action alternative and reasonable alternatives outside the agency's current legal jurisdiction, and discussed categories of alternatives it ultimately eliminated from more detailed analysis. In addition, the alternatives were responsive to the agency's purpose and need and were reasonably developed, rigorously explored, and objectively evaluated, supporting a fully informed and well-considered decision.

NEPA Requirements

NEPA section 102(2)(C)(iii) requires an EIS to include alternatives to the proposed action. 42 U.S.C. § 4332(2)(C)(iii). In implementing this requirement, CEQ regulations mandate, among other things, that agencies shall: "(a) Rigorously explore and objectively evaluate all reasonable alternatives . . . ; (b) Devote substantial treatment to each alternative considered in detail including the proposed action . . . ; (c) Include reasonable alternatives not within the jurisdiction of the lead agency; and (d) Include the no action alternative." 40 C.F.R. § 1502.14. In addition, FAA's NEPA Orders recognize that analysis of alternatives:

> "…is the heart of the EIS. It presents a comparative analysis of the no action alternative, the proposed action, and other reasonable alternatives to fulfill the purpose and need of the action . ... Reasonable alternatives not within the jurisdiction of the lead agency should be considered. The FAA may include alternatives proposed by the public or other agenc[ies]. However, they must meet the basic criteria for any alternative: it must be reasonable, feasible, and achieve the project's purpose. ...."

FAA Order 1050.1E, Section 506e. Again, courts grant substantial deference to agency actions and apply a "rule of reason" when evaluating an agency's range of alternatives. *Vermont Yankee Nuclear Power Corp. v. NRDC*, 435 U.S. 519 (1978); *NRDC v. Morton*, 458 F.2d 827 (D.C. Cir. 1972).

NEPA requires only that the agency weigh all reasonable alternatives and come to a "fully informed" and "well-considered" decision. *North Slope Borough v. Andrus*, 642 F.2d 589, 599 (D.C. Cir. 1980). *See also Vermont Yankee*, 435 U.S. at 558; *NRDC v. Hodel*, 865 F.2d 288, 294 (D.C. Cir. 1988). "This 'rule of reason governs both "*which* alternatives the agency must discuss, and the *extent to* which it must discuss them." ' " *City of Grapevine*, 17 F.3d at 1506 (citation omitted) (emphasis in original).

A project's purpose and need statement guides the appropriate range of alternatives to be considered by an agency. In turn, the chosen range of alternatives must achieve the proposed action's purpose and need, and the preferred alternative must be the one that best satisfies those parameters. *See Citizens Against Burlington*, above. It is the agency itself that must determine the alternatives to be considered and how extensive the treatment of those alternatives should be. *Vermont Yankee*, 435 U.S. at 558. Although the choice of alternatives is a matter of agency discretion, the alternatives must be "bounded by some notion of feasibility." *Id.* at 551. A court must uphold an agency's decision so long as "the alternatives are reasonable and the agency discusses them in reasonable detail." *Citizens Against Burlington*, 938 F.2d at 196.

FAA Actions

FAA began its identification of alternatives by considering five categories of potential alternatives: (1) alternate modes of transportation and communication; (2) changes in airport use, including additional infrastructure; (3) congestion management measures; (4) improved air traffic control technology; and (5) airspace redesign.[136] Of these five categories, only the airspace redesign category met FAA's purpose and need, and the agency therefore eliminated the other categories from further consideration.[137]

FAA then formed the Airspace Redesign Team working group, charged with designing and evaluating conceptual airspace redesign alternatives. The Airspace Redesign Team was composed of representatives from each of the affected FAA ATC facilities (New York TRACON, Philadelphia TRACON, New York Center, Washington Center and Boston Center), as well as representatives from ATC facilities outside the study area. In addition to internal coordination among working group representatives, FAA solicited and considered input from airlines, airport operators, and other air transportation stakeholders.[138] Recommendations also were received from RTCA, Inc., a nonprofit corporation that functions as a Federal Advisory Committee and develops industry consensus regarding air traffic management issues.[139] The Airspace Redesign Team explored four broad airspace redesign concepts—Four Corner-Post, Modifications to Existing Routing, Ocean Routing, and Clean Sheet—that met a series of nine objectives.[140]

Four basic airspace redesign alternatives were then developed for detailed consideration, ranging from the Future No Action Airspace Alternative (maintaining the status quo) to the Integrated Airspace Alternative Variation with Integrated Control Complex (creating a more fully integrated airspace). These alternatives met the purpose and need, reflected public concern about the airspace redesign, or were required by NEPA.[141] In the final EIS, FAA recommended the Integrated Airspace Alternative Variation with Integrated Control Complex (ICC) as the preferred alternative, and in its September 2007 Record of Decision, named this alternative, together with noise mitigation measures, as the selected project.[142]

FAA conducted detailed analyses of each of the four basic alternatives. The primary evaluative tool used was the Total Airspace and Airport Modeler (TAAM), a fast-time simulation tool that can model ground, terminal, and en-route airspace environments. Under FAA's instruction, MITRE's CAASD used TAAM to simulate all of the alternatives and measure their operational impacts. The area's eight airports likely to be most affected by the airspace redesign were included in the model in detail. In addition, for each alternative, FAA conducted an analysis of the noise impacts resulting from redesign-related changes in air operations.

FAA used a specific set of criteria to evaluate and compare each alternative, both quantitatively and qualitatively. Alternatives were evaluated based on two types of criteria—operational viability and operational efficiency—to determine how well each alternative met the purpose and need.[143] In its evaluation of each alternative, FAA summarized the changes that would occur at specific airports under each alternative, and then evaluated and compared each alternative using the criteria. An analysis of the noise impacts also was conducted for each alternative.[144]

Analysis

FAA's range of alternatives complied with NEPA because: (a) FAA included a no action alternative and reasonable alternatives outside the agency's jurisdiction and discussed the categories of potential alternatives eliminated from detailed analysis; (b) the alternatives were responsive to the agency's purpose and need; and (c) the alternatives were reasonably developed, rigorously explored, and objectively evaluated, which supported a fully informed and well-considered decision.

As noted, FAA considered four basic alternatives in detail, including a "no action" alternative as required by CEQ regulations.[145] Additionally, FAA considered, although eventually eliminated from detailed analysis, categories of alternatives outside the agency's jurisdiction, for instance, use of satellite airports and improvements to airport infrastructure, also as required by CEQ regulations.[146]

Further, these alternatives were responsive to the purpose and need statement, and FAA engaged in a detailed and collaborative process to ensure this result. This process included development of the Airspace Redesign Team, solicitation of input from facilities outside the study area as well as various stakeholders, and exploration of broad concepts using specific objectives. In addition, FAA used the project's purpose and need to evaluate the four alternatives both quantitatively and qualitatively. Each evaluative criterion flowed from an aspect of the stated purpose and need.

Finally, the iterative process used by FAA and the detail and consideration given to the development and exploration of the alternatives suggest that FAA's range and scope of alternatives was "well-informed" and developed and that the agency came to a "well-considered" decision. FAA engaged in a detailed process to develop and explore a range of alternatives. In creating the Airspace Redesign Team, which then collaborated with outside and industry stakeholders and RTCA, the agency sought to gather all potential input and objectively develop feasible alternatives with both a regional and national focus. Additionally, once the alternatives were winnowed down, FAA engaged in another detailed process which modeled these alternatives. FAA's last step was to evaluate the alternatives based on criteria developed to meet the project's purpose and need. This process, which took several years to complete, contributed to an objective, well-developed, and well-considered range and scope of alternatives. The exploration of the alternatives based on the criteria also was rigorous. FAA conducted quantitative and qualitative evaluations, which are discussed at length in the EIS. FAA summarized the changes that would occur at specific airports under each alternative and then evaluated the alternative using the criteria. FAA also conducted an analysis of the noise impacts for each alternative. Under the governing "rule of reason" standard, these actions produced a reasonable range of alternatives, which were rigorously explored.[147]

3. Did FAA Reasonably Decide not to Analyze Environmental Effects of the Potential Growth Resulting from its Airspace Redesign?

As discussed below, FAA complied with NEPA in deciding not to analyze environmental effects of the potential growth resulting from its airspace redesign. In the aviation context, courts have uniformly upheld FAA's decision not to analyze the effects of induced growth or induced capacity where the project's purpose was not to induce growth and the project did not entail construction of additional runway capacity.

NEPA Requirements

One of NEPA's overarching requirements is that the agency consider and discuss the environmental effects of its proposed action. 40 C.F.R. §§ 1502.16, 1508.7. This includes considering all direct and indirect effects and cumulative impacts of an action. *Id.* CEQ regulations define "indirect effects" as reasonably foreseeable effects caused by the action which appear later in time or farther in distance. 40 C.F.R. § 1508.8(b). This includes growth-inducing effects and other effects related to induced changes in the pattern of land use, population density or growth rate, and related effects on air and water and other natural systems, including ecosystems. *Id.*

FAA Actions

A number of external stakeholders involved in the New York/New Jersey/Philadelphia Airspace Redesign argued that the greater efficiencies resulting from a redesign of the airspace would necessarily and foreseeably lead to an increase in air traffic volume, and that FAA therefore was required to analyze the environmental effects of this induced traffic. FAA disagreed, stating that the airspace redesign itself—which does not include infrastructure improvements such as additional runways—would not induce growth, and thus the agency was not required to (and did not) analyze the environmental impacts of such growth. As FAA explained in its response to comments, the airspace redesign "does not increase capacity. It increases the efficiency with which existing capacity is used . . .."[148] Similar statements were part of the Administrative Record.[149]

Analysis

FAA complied with NEPA in deciding not to analyze the potential growth-inducing effects of the airspace redesign, because the purpose of the redesign is neither to induce growth, nor create additional runway capacity.[150] In cases where the project's purpose did not include growth inducement and where the project did not create additional runway capacity, courts have routinely upheld FAA's decision not to analyze the effects of induced growth.

In *Seattle Community Council Federation v. FAA*, 961 F.2d 829, 836 (9th Cir. 1992), for example, the U.S. Ninth Circuit Court of Appeals found that because planned rerouting procedures at Seattle-Tacoma International Airport were designed to enhance the safety and efficiency of existing traffic rather than to induce growth, and involved no physical infrastructure changes, NEPA did not require FAA to address the environmental effects of an increased number of flights that might indirectly result from such efficiency improvements. In upholding FAA, the court emphasized the project's stated purpose of improving safety and efficiency, the same as FAA's purpose and need here. Although the challenges in *Seattle Community Council* related to FAA's failure to prepare an EIS, rather than the adequacy of an EIS as in the airspace redesign situation, the underlying issue is the same.

Similarly, in *Morongo Band of Mission Indians v. FAA*, 161 F.3d 569 (9th Cir. 1998), FAA proposed to alter arrival procedures at Los Angeles International Airport without making changes to the airport's physical infrastructure. Relying in part on *Seattle Community Council*, the Ninth Circuit ruled that FAA complied with NEPA even though its pre-EIS environmental assessment did not consider possible growth-inducing effects of its action. The court explained that while "[g]rowth certainly may be a foreseeable indirect effect of the [project] . . ., the project was implemented in order to deal with *existing* problems; the fact that it might also facilitate further growth is insufficient to constitute a growth-inducing impact under 40 C.F.R. §1508.8(b)." *Id.* at 580 (emphasis added). And in *City of Los Angeles v. FAA*, 138 F.3d 806 (9th Cir. 1998), the court found that FAA had given the requisite "hard look" at the environmental effects of a terminal expansion project at Burbank-Glendale-Pasadena Airport that did not include additional runways, even though FAA did not consider expansion's growth-inducing effects. The court relied in part on the fact that FAA's technical forecasts of air transportation demand and airfield capacity should be accorded significant deference. *Id.* at 807, n.2.

Finally, in *City of Olmsted Falls v. FAA*, 292 F.3d 261 (D.C. Cir. 2002), the U.S. D.C. Circuit Court of Appeals rejected a challenge to FAA's plan to move an existing runway at Cleveland Hopkins International Airport as part of an airport redevelopment project. The city argued that FAA had improperly failed to consider increased air emissions that would result from increased airport capacity. The court found FAA's actions were reasonable, however, even though FAA did not consider increased capacity, because as FAA determined, air travel demand and increased capacity were independent of rerouting the runway. (As the court phrased FAA's argument, "In other words, 'if you don't build it, they will come anyway.'" *Id.* at 272.) As in *City of Los Angeles*, the court relied in part on the fact that "FAA's expertise in forecasting air transportation demand and airfield capacity are areas where courts accord significant deference." *Id.* at 272 (citations omitted).

These cases are unlike *California v. DOT*, 260 F. Supp. 2d 969 (N.D. Cal. 2003), discussed in section 1 above. There, the court found that FAA's decision not to analyze the environmental effects of a substantial growth-inducing expansion of a small town airport violated NEPA. In contrast to the present case, the expansion project in *California* would have extended and strengthened the airport's runway, created an air carrier apron, added access roads and parking, and constructed a passenger terminal complex. As noted above, the *California* court found that the very purpose of the expansion was to induce growth, by attracting thousands of additional tourists to a local ski resort. Distinguishing the projects in *Seattle Community Council and City of Los Angeles* because they each occurred in "a large city with substantial established commercial air service," the court in *California* emphasized that the project there involved "expansion of an airport to accommodate regular commercial air service, where none currently exists, in a scenic, mountain region with unique, largely undeveloped natural resources." 260 F. Supp.2d at 977. As the court concluded, "[s]uffice it to say that [FAA's] argument that the airport project is growth-accommodating rather than growth-inducing . . . is belied by the record." *Id.* at 976-77. Because FAA had failed to take a hard look at the project's environmental consequences, the court found that the agency's decision not to conduct an EIS was unreasonable, and ordered a halt to the project until an EIS was completed.

We believe FAA's decision not to analyze potential growth-inducing effects of the New York/New Jersey/Philadelphia Airspace Redesign is akin to *Seattle Community Council, City of Los Angeles*, and *City of Olmsted Falls* and distinguishable from *California v. DOT*. As discussed in section 1 above, the purpose of the airspace redesign is to increase the efficiency and reliability of the airspace system while maintaining safety. The purpose is not, as in *California v. DOT*, to induce growth or enhance runway capacity, and the redesign does not, as in *California*, involve any additional physical infrastructure. In addition, in the current case, FAA specifically found that growth would not result from the project itself, a technical judgment similar to those accorded deference by *City of Los Angeles* and *City of Olmsted Falls*. Further, the project will, as in *Seattle Community Council* and *City of Los Angeles*, affect a region already saturated with air traffic. Under existing law, therefore, FAA was not required to analyze the potential indirect effects of increased growth that may result from the redesign.

4. Did FAA Reasonably Involve the Public in the Environmental Review Process?

As discussed below, FAA complied with NEPA in reasonably involving the public in its environmental review process. FAA conducted an early and open process and made substantial efforts to involve the public throughout the environmental review process. It also sponsored a reasonable number of public meetings and provided the public with adequate advance notice. Finally, FAA considered and responded to comments in an adequate manner.

NEPA Requirements

CEQ regulations and FAA Order 1050.1E implement NEPA requirements for public participation. The regulations require that agencies involve the public in a number of different ways. An agency should begin by conducting an "...early and open process for determining the scope of issues related to a proposed action"[151] and [m]ake "diligent efforts to involve the public..."[152] throughout the environmental review process. To this end, the regulations require that an agency "provide public notice of NEPA-related . . . public meetings . . . and the availability of environmental documents..."[153] and hold or sponsor public meetings whenever appropriate or in accordance with statutory requirements...[154] so as to solicit appropriate information from the public.[155] Finally, in receiving information from the public, the regulations require an agency to "assess and consider comments both individually and collectively, and . . . respond . . . stating its response in the final statement . . .."[156]

FAA Order 1050.1E provides further guidance on how FAA should comply with the foregoing regulations. As relevant here, the order provides that "[a]t the earliest appropriate stage of the action and early in the process of preparing NEPA documentation, the responsible . . . official . . . must provide pertinent information to the affected community and agencies and consider the affected communities' opinions..."[157] To accomplish this, the order specifies that FAA must provide the public with an opportunity to review and comment on the draft EIS and must formally respond to those substantive public comments in the final EIS. [158] Additionally, the order emphasizes the discretion granted to FAA in this area by stating that there is "...no standard approach to public scoping..."[159] and "...no requirements for a scoping meeting or for a specific number of meetings";[160] nonetheless, "...it is important that FAA facilitate public participation in [the scoping process]...."[161]

FAA Actions

FAA involved the public in an extensive prescoping process for the airspace redesign. This included holding 31 prescoping public meetings, referred to as "Airspace Redesign Workshops," as a strategy to begin public education on airspace redesign, concepts, and expected benefits of the project.[162] A total of 1,174 people attended the workshops and 712 comments were submitted. Further, during the formal scoping period, FAA held 28 meetings throughout the study area with the purpose of educating the public on the purpose and need of the project. A total of 1,031 people attended the scoping meetings and 901 comments were submitted and presented in a Scoping Report. This was followed by a formal public comment period, which was extended to solicit further public participation. During the public comment period, FAA conducted 30 public meetings in the five states affected by the redesign's proposed alternatives. Subsequently, FAA issued a Noise Mitigation Report on the preferred alternative and solicited public input on the report by holding seven additional meetings, with at least one meeting in each state affected by the preferred alternative.

FAA also advertised the meetings in local media through newspaper advertisements, press releases, and public service announcements. Additionally, FAA compiled a running list of all participants at each meeting and sent any subsequent announcements to those individuals by postcard, e-mail, or other means. FAA held its initial meetings in potentially affected communities and held its draft EIS and noise-mitigation meetings in areas where the most significant impacts of the airspace redesign would be felt. These meeting locations were accessible by public transit and were located to generate the most participation. (As discussed in section 5 below, the meeting locations also satisfied environmental justice directives.) At each meeting, FAA provided handouts and visual maps and displayed videos to assist the public in understanding the process. Representatives from FAA, including air traffic controllers and contractors, were on hand at scoping, draft EIS and noise-mitigation meetings to explain the project and answer questions. Throughout the EIS process, FAA solicited public comments and responded to each substantial comment, as required by CEQ regulations.[163] Comments could be made at public meetings as well as by U.S. mail and e-mail. The public comment periods lasted at least 30 days and in some cases were extended to accommodate additional public participation.[164]

Analysis

Making a determination whether the public has been adequately involved in the EIS process is "a fact-intensive inquiry made on a case-by-case basis." *Biodiversity Conservation Alliance v. Bureau of Land Mgmt.*, 404 F. Supp. 2d 212, 220 (D.D.C. 2005). In this case, FAA conducted an early and open process and made considerable efforts thereafter to involve the public in the environmental review process. The prescoping campaign designed by the agency garnered extensive public participation and comments. The large number of meetings held in accessible areas in the study area (chosen to elicit the largest number of attendees) reflects FAA's efforts to educate the public about the redesign and address concerns, gather information, and solicit comments early in the process. FAA drew large crowds at the 31 meetings and received numerous formal public comments. Further, FAA made substantial efforts to involve the public in the formal scoping period, post-draft EIS period, and post-noise mitigation period, by similarly conducting another round of public meetings (65 in total), which educated the public on additional elements of the project. Meetings were accessible, they drew a substantial audience, and yielded a substantial number of public comments. Additionally, in each of these stages, FAA engaged in an open campaign to invite the public's attendance and comments by advertising through various media and keeping a running list of attendees. FAA also carefully chose locations which would contribute to this open process—accessibility for all and locations which would generate the most participation. Finally, FAA structured its meetings so that the public could learn about the process in various formats and ask questions of FAA and its contractors. These factors all contributed to the conduct of an "early and open process" and the use of "diligent efforts" to involve the public, as the CEQ regulations require, including sponsoring a reasonable number of public meetings (96 in all) and providing adequate notice of those meetings in various forms.

FAA solicited public comments in various forms and responded to comments consistent with its responsibilities outlined in its Order. FAA solicited comments at public meetings and accepted them via U.S. mail and e-mail, as well. Public comment periods lasted 30 days or longer, and some were extended to garner additional public participation. FAA advised that it responded to all substantive comments received in the final EIS,[165] directly addressing the question presented and showing consideration of concerns. Based on FAA's solicitation efforts and responses to substantive comments in the final EIS, FAA efforts met CEQ procedural regulations and FAA Order 1050.1E. While there is no set process for addressing public participation requirements, in our view, FAA adequately addressed the issue consistent with NEPA regulatory requirements.

5. Did FAA Reasonably Consider Environmental Justice Issues In Its Environmental Review Process?

As discussed below, in its EIS review process for the New York/New Jersey/Philadelphia Airspace Redesign, FAA satisfied the key environmental justice directives contained in Executive Order 12898, CEQ guidance documents, and DOT Orders. FAA prepared an analysis that identified minority and low-income populations significantly impacted by the proposed redesign and determined whether the impact on these populations was disproportionate. FAA also involved these individuals throughout the environmental review process. In addition, FAA mitigated these impacts through alteration of the preferred alternative, which included reducing the number of departure dispersal headings, shifting certain departure routes, and changing certain arrival altitudes.

Executive Order 12898 and Other Requirements

Executive Order 12898, "Federal Actions to Address Environmental Justice in Minority Populations and Low-Income Populations," provides that "...each Federal agency shall make achieving environmental justice part of its mission by identifying and addressing . . . disproportionately high and adverse human health or environmental effects of its programs, policies, and activities on minority populations and low-income populations...." *Id.* Sec. 1-101. The Presidential memorandum accompanying issuance of the Executive Order emphasized the importance of considering these effects when an environmental analysis is required under NEPA. *See* "Memorandum For the Heads of All Departments and Agencies, Re: Executive Order on Federal Actions to Address Environmental Justice in Minority Populations and Low-Income Populations," Comprehensive Presidential Documents No. 279 (Feb. 11, 1994). CEQ, which oversees federal agency compliance with Executive Order 12898, has developed guidance to assist agencies in meeting their responsibilities under the order. *See Environmental Justice: Guidance Under the National Environmental Policy Act* (Dec. 10, 1997) (hereafter, CEQ Guidance). DOT also has issued an order to guide implementation of the Executive Order, which sets broad environmental goals.[166]

By their terms, the Executive Order and implementing documents provide agencies with considerable flexibility in seeking to achieve these environmental justice goals. The Executive Order, for example, states that agencies shall make environmental justice part of their missions "[t]o the greatest extent practicable and permitted by law...[and]...as appropriate . . .." The CEQ Guidance describes its terms as flexible and serving as "...a point of departure, rather than a conclusive direction in applying the terms of the executive order,[167] and recognizes "there is [no] standard formula for how environmental justice issues should be identified or addressed."[168] There also is not a particular formula by which to address these in an EIS; agencies simply must present such concerns in a manner that is "...clear, concise, and comprehensible...."[169]

Against this backdrop of flexibility, Executive Order 12898 and these related documents direct agencies to make efforts to ensure the involvement of low-income and minority populations as part of the agency's consideration of the environmental impacts of its actions. The CEQ Guidance, for example, states that "...agencies should assure meaningful community representation . . . [and] endeavor to have complete representation of the community as a whole . . . as early as possible if it is to be meaningful."[170] To this end, an agency should "...provide opportunities for effective community participation in the NEPA process, including . . . improving the accessibility of public meetings, crucial documents, and notices"[171] and "develop effective public participation strategies, . . . acknowledge and seek to overcome linguistic, cultural, institutional, geographic, and other barriers to meaningful participation, and should incorporate active outreach to affected groups."[172]

The CEQ Guidance recommends consideration of these impacts at various stages of the environmental review process. In the scoping process, the CEQ Guidance suggests an agency "seek input from low income populations [and] minority populations . . . as early in the process as information becomes available"[173] by use of "...local resources, community and other non-governmental organizations, and locally targeted media."[174] The following are cited as examples of ways to help inform the public during the scoping process: providing "a description of the proposed action," "maps, drawings, and other appropriate material or references," "an agency point of contact," and "timely notice of locations where comments will be received or public meetings held."[175]

Further, to solicit public participation, the CEQ Guidance recommends "...adaptive or innovative approaches to overcome linguistic, institutional, cultural, economic, historical, or other potential barriers to effective participation . . .."[176] Suggestions for developing a public participation strategy include: "translation of major documents...";[177] "...opportunities for limited-English speaking members of the affected public to provide comments throughout the NEPA process;"[178] use of different media (e.g., size, formats);[179] and "use of locations and facilities that are local, convenient, and accessible to . . . low-income and minority communities . . .."[180] Finally, when analyzing alternatives, the CEQ Guidance suggests an agency consider "...the views it has received from the affected communities, . . . the magnitude of environmental impacts associated with alternatives that have a less disproportionate and adverse effect on low-income populations [and] minority populations" and "...community views in developing and implementing mitigation strategies."[181]

FAA Actions

Throughout its environmental review process for the New York/New Jersey/Philadelphia Airspace Redesign, FAA sought to ensure the involvement of low-income and minority populations who would be significantly affected by the redesign. FAA conducted an environmental justice assessment to identify where those populations were located,[182] and determined that minority populations near LaGuardia, Newark, and Philadelphia would experience significant noise impacts under what ultimately became the preferred alternative. Beginning in the prescoping phase, therefore, and continuing through the noise-mitigation report phase, FAA held public meetings about the redesign in areas accessible by public transit and in low-income and minority communities, such as Newark, N.J., with disproportionately high and adverse noise impacts. FAA provided translators at these meetings to facilitate communication about the project and used different forms of media—maps, printouts, and video presentations, in addition to live speakers—to convey the elements and effects of the project. The meetings were publicized by advertisements in multiple publications, including local foreign language media, and by outreach to church leaders and community organizations. The public could submit comments to FAA at these targeted meetings. All of these measures were in addition to FAA's general public information and public comment opportunities for the redesign, as discussed in section 4 above.

FAA concluded that there would be a disproportionate impact on minority populations near LaGuardia and Newark; however, with mitigation measures, these impacts would be eliminated. In addition, FAA concluded that there would be significant (though not disproportionate) impacts on minority populations near Philadelphia, but these would be eliminated once the redesign was fully implemented by 2012.[183] FAA also considered another alternative—the Ocean Routing Alternative—that would have eliminated noise impacts on these populations altogether but, as noted, that alternative was not selected because it did not meet the redesign's purpose and need.

Analysis

FAA satisfied the environmental justice requirements of Executive Order 12898 and the related CEQ and FAA regulations and guidance. As described in the final EIS, FAA conducted an assessment to identify minority and low-income populations that would be significantly affected by the redesign, determined whether the impact on these populations was disproportionate, and involved those populations throughout the environmental review process with sensitivity to issues that could affect their level of participation. As Executive Order 12898 and CEQ guidance require, FAA engaged in reasonable initiatives to ensure these populations had the chance to participate in the review process, for example, by holding meetings in areas accessible by public transit and in those communities most significantly affected by the project. FAA also adhered to directives to overcome cultural and linguistic barriers and involve community and nongovernmental organizations in the notification process, by publicizing public meetings in various publications in both English and foreign language formats. When conducting these meetings, FAA heeded the directive to convey project information in various formats, by using maps, printouts, video, and speakers, as well as the Internet. These efforts generated a large number of comments from the affected populations, and FAA adequately responded to each substantive comment in the final EIS.

Finally, FAA considered these populations in developing its alternatives, as the CEQ Guidance requires, by considering potential noise impacts on these communities and considering the Ocean Routing Alternative that would have eliminated disproportionately high and adverse noise impacts. FAA also mitigated the preferred alternative, thereby eliminating significant noise impacts no later than full implementation, in compliance with the CEQ Guidance requiring consideration of community views in developing and implementing mitigation strategies. These efforts and activities met the requirements of Executive Order 12898, CEQ Guidance, and the other directives "[t]o the greatest extent practicable and permitted by law…" and "as appropriate…."

APPENDIX III: OPERATIONAL COMPARISON OF ALTERNATIVES

Purpose and Need Evaluation Criteria	How Measured	Alternative				
		Future No Action	Modifications to Existing Airspace	Ocean Routing Airspace	Integrated Airspace without ICC	Integrated Airspace with ICC
Reduce complexity	Jet route delays + time below 18,000 feet (minutes)	12	12	12	11	10
	Arrival distance below 18,000 ft (nautical miles)	96	95	99	96	102
Reduce voice communications	Max interfacility hand-offs per hour	525	525	521	529	382
Reduce delay	Traffic weighted arrival delay 2011 (minutes)	22.9	22.6	23.6	22.8	19.9
	Traffic weighted departure delay 2011 (minutes)	23.3	20.9	29.5	20.8	19.2
Balance controller workload	Equity of west gate fix traffic counts	0.37	0.37	0.37	0.34	0.30
Meet system demands and improve user access to system	End of day's last arrival push (time)	23:54	23:54	23:54	23:54	23:00
Expedite arrival and departures	Time below 18,000 ft (minutes)	18.5	18.2	18.8	18.2	18.6
	Change in route length per flight (nautical miles)[a]	0.0	0.0	4.5	-1.2	3.7
	Change in block time (minutes per flight)[a]	0.0	-0.9	3.9	-1.0	-1.4
Flexibility in routing	Delay saved per flight per day	0.0	0.0	0.0	0.0	12.6
Maintain airport throughput	Arrival max sustainable throughputs	223	223	223	223	238
	Departure max sustainable throughputs	238	239	221	240	245

Source: FAA.

[a] A negative value indicates a new decrease in the category.

Appendix IV: Summary of the Integrated Airspace Alternative with ICC

Airport	Changes from the Future No Action Alternative
JFK	North departure gate shifted 15 miles northeast
	New distant procedures for aircraft heading to the North departure gate
	West departure gate extended to the north and to the south
	New procedures for aircraft heading to the West departure gate
	Future No Action Ocean departure gate split into Ocean and South departure gates
	New distant procedures for aircraft heading to the Ocean departure gate
	New procedures for aircraft heading to the South departure gate
	North arrival post shifted 5 miles southeast
	New distant procedures for aircraft arriving from the North arrival post
	East arrival post shifted northwest
	New procedures for aircraft arriving from the East arrival post
	South arrival post shifted to the northeast
	New procedures for aircraft arriving from the South arrival post
LaGuardia	East departure gate shifted east
	North departure gate shifted 15 miles northeast
	New procedures for aircraft heading to the North departure gate
	West departure gate extended to the north and to the south
	New procedures for aircraft heading to the West departure gate
	South departure gate shifted to the northwest
	New procedures for aircraft heading to the South departure gate
	North arrival post shifted 30 miles east
	New procedures for aircraft arriving from the North arrival post
	West arrival posts shifts to coincide with Future No Action South arrival post
	New procedures for aircraft arriving from the west to coincide with the South arrival post
	West arrival flow split into two arrival flows, one to the north and one to the south
	New departure headings for aircraft departing Runway 4 to the North departure gate
	New departure headings for aircraft departing Runway 4 to the East departure gate
Newark	New departure headings for all runways and all gates
	East departure gate shifted to the east
	New procedures for aircraft heading to East departure gate
	North departure gate shifted to the northeast
	New procedures for aircraft heading to the North departure gate
	West departure gate expanded to the north and south
	New procedures for aircraft heading to the West departure gate
	South departure gate shifted to the southwest
	New procedures for aircraft heading to the South departure gate
	New Ocean departure gate
	New procedures for aircraft heading to the Ocean departure gate
	North arrival post moved to 50 miles north of Newark airport
	New procedures for aircraft arriving from the North arrival post

Airport	Changes from the Future No Action Alternative
	West arrival post shifted to be near Greenville, New York
	West arrival flow split into two arrival flows, one to the north and one to the south
	New procedures for aircraft arriving from the South arrival post
	Use of both parallel runways for arrivals
Teterboro	Departure gates match those of Newark Integrated Airspace with ICC
	New distant procedures for aircraft heading to the North departure gate
	New distant procedures for aircraft heading to the West departure gate
	New distant procedures for aircraft heading to the South departure gate
	West arrival post shifted 15 miles south
	New procedures for aircraft arriving from the West arrival post
	New procedures for aircraft arriving from the West arrival post from the vicinity of Yardley, Pennsylvania
Westchester County Airport	North departure gate shifted 15 miles northeast
	New distant procedures for aircraft heading to the North departure gate
	West departure gates extended to the north and to the south
	New procedures for aircraft heading to the West departure gate
	South departure gate shifted to the west
	New departure procedures for aircraft departing to the South gate
	North arrival post shifted to the east
	New distant procedures for aircraft arriving from the North gate
	New distant procedures for aircraft arriving from the south
Philadelphia	West departure gate expanded to the northwest
	New procedures for aircraft heading to the West departure gate
	East departure gate is shifted to the east
	New procedures for aircraft heading to the East departure gate
	West arrival post shifts to the northeast
	New distant procedures for aircraft arriving from the West arrival post
	New departure headings for aircraft heading to the North, East, West, Southwest, and South departure gates
	Additional route added to North arrival post

Source: FAA.

Appendix V: Comments from the Department of Transportation

U.S. Department of Transportation
Office of the Secretary of Transportation

Assistant Secretary for Administration

1200 New Jersey Avenue, SE
Washington, DC 20590

June 19, 2008

Ms. Susan Fleming
Director, Physical Infrastructure Issues
U.S. Government Accountability Office
441 G Street, NW
Washington, DC 20458

Dear Ms. Fleming,

Representatives from the Federal Aviation Administration (FAA) and the Office of the Secretary have reviewed the GAO draft report, "*FAA Airspace Redesign: An Analysis of the New York/New Jersey/Philadelphia Project.*" While we are pleased to see that the GAO review concluded the actions taken by FAA with regard to the National Environmental Policy Act were reasonable, the Department does not necessarily agree with everything in the draft report. However, we are constrained from offering a detailed analysis of the draft report at this time, as the matters it covers are the subject of pending litigation. Specifically, there are currently 12 consolidated lawsuits before the U.S. Court of Appeals for the District of Columbia Circuit that challenge one or more aspects of the airspace redesign project.

Neither the 73 Petitioners, nor the Federal Respondents, have yet submitted briefs on the merits. Therefore substantive comments with regard to the GAO draft report could potentially prejudice our ability to defend this project in litigation.

We appreciate the opportunity to review the draft report. Please contact Mr. Martin Gertel, Director of Audit Relations, on (202) 366-5145 with any questions.

Sincerely,

Linda J. Washington

END NOTES

[1] The five major airports include New York's John F. Kennedy International Airport (JFK) and LaGuardia Airport (LaGuardia), New Jersey's Newark Liberty International Airport (Newark) and Teterboro Airport (Teterboro), and Pennsylvania's Philadelphia International Airport (Philadelphia).

[2] Airspace for this project is the navigable area used by aircraft for purposes of flight. In addition, FAA maintains infrastructure for the National Airspace System to include FAA air traffic control (ATC) system, which relies on an extensive array of information technology systems, including radars, automated data processing, navigation, and communication equipment; and ATC facilities.

[3] In 2006, over 100 million passengers used JFK, LaGuardia, and Newark, alone.

[4] The estimated benefits are based on FAA's operational analysis for 2011; however, full implementation is not estimated to be completed until 2012.

[5] Twelve of the 13 lawsuits are based primarily on NEPA grounds and have been consolidated in the U.S. Court of Appeals for the D.C. Circuit under the lead case, *County of Rockland v. FAA*, No. 07-1363. The other lawsuit, *County of Delaware v. USDOT*, No. 07-1385 (D.C. Cir.), challenges the redesign based on FAA's allegedly improper implementation of the Clean Air Act.

[6] The court in the NEPA litigation has denied requests that FAA temporarily stop implementation while the lawsuits are being resolved.

[7] The three similar FAA-sponsored redesigns included the Expanded East Coast Plan (1995), Chicago Terminal Airspace Project (2001), and Potomac Consolidated TRACON Facility Airspace Redesign (2002). These were determined to be most similar to the New York/New Jersey/Philadelphia Airspace Redesign project because they all included an EIS, changes to air traffic routes involving multiple airports, and an analysis of noise impacts.

[8] In this report, the term "contractors" refers to organizations used to conduct the operational and environmental analyses. This includes MITRE, the organization that was responsible for conducting the operational analysis. MITRE is a not-for-profit corporation, which operates a Federally Funded Research and Development Center (FFRDC) sponsored by FAA—the Center for Advanced Aviation System Development. An FFRDC is a unique organization that assists the U.S. government with scientific research and analysis, development and acquisition, and systems engineering and integration. FFRDCs are charged with addressing long-term problems of considerable complexity, analyzing technical questions with a high degree of objectivity, and providing creative and cost-effective solutions to government problems.

[9] Although we took stakeholder concerns expressed after the NEPA process into account when developing our five key issues, a court may consider a petitioner to have waived their right to challenge issues that were not raised during the process.

[10] For this airspace redesign project, FAA included in its study area 16 satellite airports, including (1) Allentown/Lehigh Valley International, (2) Atlantic City International, (3) Bridgeport/Igor I. Sikorsky Memorial, (4) Caldwell/Essex County, (5) Westhampton Beach/The Francis S. Gabreski, (6) Islip Long Island MacArthur, (7) Linden, (8) Morristown Municipal, (9) Newburgh/Stewart International, (10) New Haven/Tweed-New Haven, (11) Northeast Philadelphia, (12) Republic, (13) Trenton/Mercer County, (14) White Plains/Westchester County, (15) Wilmington/New Castle County, and (16) McGuire Air Force Base. There are other airports physically located within the region, but they were not included in FAA's operational analysis because they do not have a significant amount of Instrument Flight Rule traffic, thus there would be little or no change to their operations as a result of the proposed action. Instrument Flight Rule traffic from these other airports was included in the environmental analysis as overflights.

[11] For example, JetBlue Airlines uses JFK as its hub airport, U.S. Airways uses Philadelphia as its hub airport, and Continental Airlines uses Newark as one of its hub airports.

[12] To minimize congestion and reduce flight delays, FAA promulgated the High Density Rule in 1968, 14 C.F.R. § 93, Subpart K, which set limits on the number of take-offs or landings—referred to as "slots"—that can occur during certain periods of the day at four congested airports—Chicago's O'Hare, Ronald Reagan Washington National, and New York's JFK and LaGuardia.

[13] Wendell H. Ford Aviation Investment and Reform Act for the 21st Century, also referred to as AIR-21, Pub. L. No. 106-181, 114 Stat. 108, Section 231, April 5, 2000.

[14] In the mid-1980s, a plan to improve airspace efficiency in and around the New York metropolitan area was implemented. The Expanded East Coast Plan focused on developing common departure routes out of the New York metropolitan area. The plan, which included changes to routes and procedures above 3,000 feet, was implemented in phases in 1987 and 1988. Based on the public reaction to the Expanded East Coast Plan, FAA recognized that the New York/New Jersey/Philadelphia Airspace Redesign had the potential to be controversial based on potential environmental impacts.

[15] The National Airspace Redesign Strategic Management Plan was developed collaboratively by local facility and regional air traffic specialists, Air Transport Association representatives, and the National Air Traffic Controllers Association (NATCA). The plan works in conjunction with the March 16, 2001, FAA/NATCA

Memorandum of Understanding regarding the National Airspace Redesign. In 2005, when the National Airspace Redesign transitioned to the Airspace Management Program, FAA terminated the assignments for NATCA representatives to the National Airspace Redesign.

[16] There are two versions of the Handbook. The Airspace Management Handbook Version 1 was first published in 1999 and consists of three parts—the Airspace Management Handbook: Checklist; the Airspace Management Handbook: Guidelines; and the Airspace Management Handbook: Metrics. In December 2005, FAA published Version 2.2, which combines all of the content of the previous documents and includes supporting information for safety risk management considerations. The steps established in these handbooks include (1) characterizing the extent of the airspace problem, (2) performing an initial evaluation (scoping), (3) initiating an airspace study, (4) conducting an airspace study, (5) summarizing and presenting the results of the airspace study, (6) implementing the recommended changes, and (7) evaluating the final implementation of the airspace redesign to ensure changes made to the airspace accomplished the redesign's intended goal. Because the FAA airspace redesign project was initiated in 1998, we are basing our review on Version 1.

[17] 42 U.S.C. §§ 4321 *et seq.*

[18] NEPA section 102(c), 42 U.S.C. § 4332(c). An EIS is not required for all federal actions. Actions that do not individually or cumulatively have a significant effect on the human environment may be categorically excluded. 40 C.F.R. § 1508.4. Additionally, an EIS is not required for projects for which a project-specific Environmental Assessment discloses no significant impact. 40 C.F.R. § 1508.9.

[19] 40 C.F.R. § 1506.6.

[20] At the start of the airspace redesign in 1998, FAA Order 1050.1D, *Environmental Impacts: Policies and Procedures,* was the current guidance. The Order was updated to FAA Order 1050.1E, *Environmental Impacts: Policies and Procedures,* on June 8, 2004, and again on March 20, 2006, to FAA Order 1050.1E, Chg 1.

[21] Executive Order 12898, *Federal Actions to Address Environmental Justice in Minority Populations and Low-Income Populations*, 59 Fed. Reg. 7629 (Feb. 16, 1994).

[22] DOT Order 5610.2, *Environmental Justice in Minority and Low-Income Populations.*

[23] Executive Order 12898, Sec. 1-101.

[24] In this airspace redesign, ICC refers to the existence of a common automation platform in existing facilities, a new facility, or a consolidated facility. A common automation platform refers to a single radar data processing system and the information it provides to controllers. It includes shared displays on screens, radar, data processing and presentation, and communications.

[25] FAA issued a corrected version of the Record of Decision on Sept. 28, 2007 after FAA identified several items in the original document that were omitted or incorrect due to editing mistakes.

[26] 5 U.S.C. § 706(2)(A).

[27] Draft EIS, p. 2-2; final EIS, p. 2-2.

[28] 40 C.F.R. § 1502.13.

[29] FAA Order 1050.1E, section 506(d).

[30] Experts we interviewed also agreed that the scope of the purpose and need was reasonable.

[31] 40 C.F.R. § 1502.14(d).

[32] 40 C.F.R. § 1502.14(a).

[33] An agency's discussion of alternatives will be upheld as long as the alternatives are reasonable and are discussed in reasonable detail. *Citizens Against Burlington, Inc. v. Busey*, 938 F.2d 190, 196 (D.C. Cir. 1991), cert. denied, 502 U.S. 994 (1991).

[34] *See Citizens Against Burlington, Inc.*, 938 F.2d at 196.

[35] In addition, experts interviewed by GAO agreed that the alternatives were feasible and met the purpose and need of the project.

[36] *See Citizens Against Burlington, Inc.*, 938 F.2d at 196.

[37] 40 C.F.R. § 1502.14(a).

[38] 40 C.F.R. § 1502.14(b).

[39] 40 C.F.R. §§ 1502.16, 1508.7.

[40] 40 C.F.R. § 1508.8(b).

[41] *Morongo Band of Mission Indians v. FAA*, 161 F.3d 569, 580 (9th Cir. 1998); *see also Seattle Community Council Federation v. FAA*, 961 F.2d 829 (9th Cir. 1992) (re-routing of flights at Seattle-Tacoma International Airport involving no infrastructure changes did not require FAA analysis of environmental effects of increased flights that may indirectly result from efficiency improvements).

[42] 40 C.F.R. § 1506.6(a).

[43] 40 C.F.R. § 1501.7.

[44] 40 C.F.R. § 1506.6(b) and (c).

[45] 40 C.F.R. § 1506.6(d), 40 C.F.R. § 1503.1(4) and 40 C.F.R. § 1503.4.

[46] FAA stated that its decision to hold prescoping meetings was due to prior experiences with airspace redesign and a desire to better educate the public on airspace redesigns.

[47] FAA's contractors stated the mailing list kept the same names throughout the various EIS periods and added new names and addresses to the master list during the process.

[48] FAA stated that the mailing list developed during the Expanded East Coast Project was used to contact members of the public about the airspace redesign.

[49] Experts we interviewed also stated that FAA's involvement of the public throughout the EIS process was sufficient given the scope and complexity of the redesign.

[50] 40 C.F.R. §§ 1503.1(4), 1503.4, 1506.6(d).

[51] Executive Order 12898, *Federal Actions to Address Environmental Justice in Minority Populations and Low-Income Populations*, Feb. 11, 1994.

[52] CEQ, *Environmental Justice: Guidance under the National Environmental Policy Act*, Dec. 10, 1997.

[53] DOT 5610.2, *Environmental Justice in Minority and Low-Income Populations.*

[54] Congressional members and citizens raised concerns about FAA's environmental justice assessment around Newark Liberty Airport during the public comment period for the draft EIS.

[55] In the final EIS, FAA stated that the definitions of "minority" and "low-income" populations in Order DOT 5610.2 were broad. Therefore the agency also used definitions from the CEQ to more narrowly define the two populations.

[56] GAO, *Surface Transportation: Many Factors Affect Investment Decisions*, GAO-04-744 (Washington, D.C.: June 30, 2004); GAO, *Highway and Transit Investments: Options for Improving Information on Projects' Benefits and Costs and Increasing Accountability for Results*, GAO-05-172 (Washington, D.C.: Jan. 24, 2005); and GAO, *Intercity Passenger Rail: National Policy and Strategies Needed to Maximize Public Benefits from Federal Expenditures*, GAO-07-15 (Washington, D.C.: Nov. 13, 2006).

[57] GAO, *Bureau of Reclamation: An Assessment of the Environmental Impact Statement on the Operations of the Glen Canyon Dam*, GAO/RCED-97-12 (Washington, D.C.: Oct. 2, 1996); GAO, *Army Corps of Engineers: An Assessment of the Draft Environmental Impact Statement of the Lower Snake River Dams*, GAO/RCED-00-186 (Washington, D.C.: July 24, 2000); GAO, *Oregon Inlet Jetty Project: Environmental and Economic Concerns Still Need to Be Resolved*, GAO-02-803 (Washington, D.C.: Sept. 30, 2002); GAO, *Assigning Air Traffic Control Costs to Users: Elements of FAA's Methodology Are Generally Consistent with Standards but Certain Assumptions and Methods Need Additional Support*, GAO-08-76 (Washington, D.C.: Oct. 19, 2007).

[58] FAA's airspace redesign team initially included air traffic controllers in the study area and neighboring areas.

[59] The Future No Action Alternative was included in the evaluation as required by NEPA and CEQ regulations. The Ocean Routing Airspace Alternative was developed by the New Jersey Citizens for Environmental Research, Inc., at the request of the New Jersey Coalition Against Aircraft Noise (NJCAAN) and was evaluated as a potential alternative due to long-standing concerns of the NJCAAN.

[60] As defined in FAA's Handbook, metrics are parameters, algorithms, or formulas used to quantify system performance, and they are measured either directly in the National Airspace System or in models of the National Airspace System.

[61] Flight track dispersion occurs because flights do not all follow the same exact path. There is natural variation of flight tracks based on operational factors, such as pilot variation and weather conditions.

[62] Use of this metric also has been upheld by courts. *See City of Grapevine v. DOT*, 17 F.3d 1502, 1508 (D.C. Cir. 1994); *Sierra Club v. DOT*, 753 F.2d 120, 128 (D.C. Cir. 1985).

[63] FICON, *Federal Interagency Review of Selected Airport Noise Analysis Issues*, August 1992.

[64] FICON was the predecessor of the Federal Interagency Committee on Aviation Noise.

[65] See GAO-04-744, GAO-05-172, and GAO-07-15.

[66] The flight operations forecast is based on the traffic demand forecast.

[67] See "FAA Airport Benefit-Cost Analysis Guidance," Office of Aviation Policy and Plans, FAA, Dec. 15, 1999. FAA officials told GAO that this guidance does not apply to airspace redesign projects. Nonetheless, the officials agreed that the economic principles embodied in the guidance are generally applicable.

[68] Total price includes air fare and travel time.

[69] More specifically, inelastic demand means that a 1 percent change in price results in a less-than-proportionate change in demand, and elastic demand means that a 1 percent change in price results in a greater-than-proportionate change in demand. By holding traffic demand constant across alternatives, FAA essentially assumed that demand is perfectly inelastic or completely unresponsive to price changes

[70] FAA's guidance also states that investments that lower average delay at an airport will generally induce some customers who formerly avoided the airport to use it, thereby placing new demands on the facility and eroding some delay savings. The guidance also states that if the project reduces delay in excess of an average of 1 minute per operation, then travel demand forecasts should be adjusted. The guidance states that as a rule of thumb, one 2 percent increment should be simulated for each 3-minute savings. In addition, the guidance states that it is DOT policy that passengers recognize small time savings and can use them effectively. Therefore, a relatively small time savings should be valued at the same rate as longer time savings.

[71] See GAO-07-15.

[72] National Airspace System Capital Investment Plan FY 2008–2012.

[73] RNAV enables aircraft to fly on any desired flight path within the coverage of ground- or-space-based navigation aids, within the limits of the capability of the self-contained systems, or a combination of both capabilities. As such, RNAV aircraft have better access and flexibility for point-to-point operations.

[74] FAA's 2006 Roadmap for Performance-Based Navigation.

[75] For example, typically route procedures that are designated with an accuracy value of 2 (designated RNAV-2), require aircraft to be within 2 nautical miles of their assigned flight path 95 percent of the flight time.

[76] These metrics also can be calculated using the same noise modeling tools used to calculate DNL. The noise model is capable of calculating up to 13 noise metrics, some of which include the supplemental metrics mentioned by the experts.

[77] Sound Exposure Level, or SEL, metric is used to measure noise exposure for a single aircraft flyover. It identifies the cumulative sound that a person is exposed to during the event if the sound were compressed into 1 second of time.

[78] The Time-Above metric can identify how much time during a designated time period—such as a day—the noise exposure levels will exceed a specified decibel level. The sound level must be specified—for example, 60 decibels. This method can then determine the length of time during a 24-hour period that noise levels will exceed 60 decibels.

[79] According to FICON.

[80] FAA calculated supplemental metrics in the Potomac Consolidated TRACON Facility Airspace Redesign and the Chicago Terminal Airspace Project.

[81] For example, a key input into the operational analysis was the travel demand forecast—an inherently uncertain input because future conditions cannot be precisely predicted.

[82] Airspace Management Handbook, Version 1, p. 21.

[83] DOT, FAA, Office of System Capacity, 2001 Aviation Capacity Enhancement Plan, December 2001.

[84] An expert interviewed said that this level of difference appeared reasonable.

[85] Specifically, the forecast underestimated airlines' use of regional jets and the level of air traffic at many regional airports.

[86] A common automation platform refers to a single radar data processing system and the information it provides to controllers. It includes shared displays on screens, radar, data processing and presentation, and communications.

[87] FAA officials do not currently know what the costs will be to develop the common automation platform.

[88] App. C of the final EIS, p. 10-4.

[89] Even though reduction of noise was not part of the purpose and need for the airspace redesign, thus not a criterion in the decision-making process, FAA emphasized that noise impacts were considered. Specifically, FAA explained that it considered noise after selecting the preferred alternative by identifying measures to mitigate the noise impacts associated with the preferred alternative. Several noise mitigation measures were incorporated into the preferred alternative, which according to FAA, would result in the mitigation of all significant noise impacts by 2011. As determined by FAA, "significant" noise impacts occur if a proposed action would result in a minimum 1.5 DNL increase where noise exposure already exceeds 65 DNL or would exceed 65 DNL after the increase. FAA also explained that it did not use noise metrics when comparing alternatives because there are no established standards for significance of noise reduction. For instance, there is no standard to answer the question, "Which is better: reducing the noise impacts for 1,000 people by 3 decibels, or reducing the noise impacts for 500 people by 5 decibels?"

[90] These methods include hedonic price techniques, which can be used to measure the effect of economic attributes like noise on housing values.

[91] FAA also did some comparative analyses that involved comparing some elements of the 2006 forecast with the observed levels in 2005 and some elements of the 2011 forecast for Philadelphia traffic with the 2012 forecast used in the Capacity Enhancement Plan.

[92] FAA *Regional Air Service Demand Study*, Summary Report. Sponsored by FAA, The Port Authority of New York and New Jersey, New York State Department of Transportation, and Delaware Valley Regional Planning Commission, May 2007.

[93] Handbook, p. 21. FAA's Benefit-Cost Analysis Guidance also recommends the use of sensitivity analysis.

[94] Experts noted that a sensitivity analysis would have less of an effect on a noise analysis than on an operational analysis. As one expert noted, this is because traffic on one day will not significantly impact the noise model, which reflects annual average conditions. In addition, according to this expert, there would need to be a 35 percent to 40 percent error in the traffic forecast to affect DNL.

[95] Although not required, a benefit-cost analysis is permitted under CEQ regulations. Under 40 C.F.R. § 1502.23,. "[i]f a cost-benefit analysis relevant to the choice among environmentally different alternatives is being considered for the proposed action, it shall be incorporated by reference or appended to the [EIS] as an aid in evaluating the environmental consequences." The regulation further states that when a cost-benefit analysis is prepared, the EIS should discuss the relationship between that analysis and any analyses of unquantified environmental impacts, values, and amenities. The regulation states that the merits and drawbacks of the

alternatives need not be displayed in a monetary cost-benefit analysis and should not be when there are important qualitative considerations.

[96] See GAO-04-744 and GAO-05-172.

[97] *FAA Airport Benefit-Cost Analysis Guidance*, Office of Aviation Policy and Plans, FAA, Dec. 15, 1999 and FAA Web page entitled "The FAA Acquisition System Toolset, AMS Policy vs. Guidance," http://fast.faa.gov (accessed March 15, 2008). FAA, along with the U.S. National Aeronautics and Space Administration Transport Canada, also is sponsoring research to develop analytical tools to assess the benefits and costs of different environmental policies and research and development investment strategies. One area of research is focusing on how to monetize the health and welfare impacts of aviation noise, local air quality, and climate effects to enable a robust cost-benefit analysis of policy alternatives.

[98] We recognize that the guidance does not apply to the Air Traffic Organization, which includes the FAA division responsible for airspace redesigns.

[99] FAA reported in the Record of Decision that the airspace redesign would reduce annual operating costs by $248 million and severe weather delay costs by another $37 million, totaling $285 million. However, in a letter to Senator Robert Menendez in February 2008, FAA estimated that there would be $300 million annual savings in direct operating costs. FAA stated that the higher reduction in annual operating costs was due to an increase in jet fuel prices.

[100] *Economic Values for FAA Investment and Regulatory Decisions, A Guide*. Prepared by GRA Incorporated for the FAA Office of Aviation Policy and Plans (Washington, D.C., Dec. 31, 2004).

[101] GAO, *Results-Oriented Cultures: Implementation Steps to Assist Mergers and Organizational Transformations*, GAO-03-669 (Washington, D.C.: July 2, 2003); GAO, *Intercity Passenger Rail: Issues Associated with the Recent Settlement between Amtrak and the Consortium of Bombardier and Alstom*, GAO-05-152 (Washington, D.C.: Dec. 1, 2004); GAO, *United States Coast Guard: Improvements Needed in Management and Oversight of Rescue System Acquisition*, GAO-06-623 (Washington, D.C.: May 31, 2006); GAO, *Telecommunications: Full Adoption of Sound Transition Planning Practices by GSA and Selected Agencies Could Improve Planning Efforts*, GAO-06-476 (Washington, D.C.: June 6, 2006); GAO, *2010 Census: Costs and Risks Must be Closely Monitored and Evaluated with Mitigation Plans in Place*, GAO-06-822T (Washington, D.C.: June 6, 2006).

[102] See GAO-03-669.

[103] See GAO-05-152.

[104] GAO, *Next Generation Air Transportation System: Progress and Challenges Associated with the Transformation of the National Airspace System*, GAO-07-25 (Washington, D.C.: Nov. 13, 2006). 105GAO, *Federal Aviation Administration: Challenges Facing the Agency in Fiscal Year 2009 and Beyond*, GAO-08-460T (Washington, D.C.: Feb. 7, 2008).

[106] See GAO-05-152, GAO-08-76, and GAO, *Next Generation Air Transportation System: Status of the Transition to the Future Air Traffic Control System*, GAO-07-784T (Washington, D.C.: May 9, 2007).

[107] See GAO-08-460T and GAO, *National Airspace System: Transformation Will Require Cultural Change, Balanced Funding Priorities, and Use of All Available Management Tools*, GAO-06-154 (Washington, D.C.: Oct. 14, 2005).

[108] The strategy is already used by some federal, state, and local agencies, such as the South Florida Ecosystem Restoration Task Force.

[109] The guide does not establish new requirements and does not constitute formal CEQ guidance.

[110] GAO, *Yellowstone Bison: Interagency Plan and Agencies' Management Need Improvement to Better Address Bison-Cattle Brucellosis Controversy*, *GAO-08-291* (Washington, D.C.: Mar. 7, 2008).

[111] See "Environmental Management Systems and NEPA Adaptive Management" available at http://www.faa.gov/regulations_policies/policy_guidance/envir_policy (accessed on May 29, 2008).

[112] The three similar FAA-sponsored redesigns included the Expanded East Coast Plan (1995), Chicago Terminal Airspace Project (2001), and Potomac Consolidated Terminal Radar Approach Control (TRACON) Facility Airspace Redesign (2002). These were determined to be most similar to the New York/New Jersey/Philadelphia Airspace Redesign project because they all included an EIS analysis, changes to air traffic routes involving multiple airports, and analysis of noise impacts.

[113] Although we took stakeholder concerns expressed after the NEPA process into account when developing our five key issues, a court may consider a petitioner to have waived their right to challenge issues that were not raised during the process.

[114] In conducting our analysis, we relied on FAA's draft and final Environmental Impact Statements and its Record of Decision, and the Administrative Record filed by FAA with the D.C. Circuit in December 2007 in the *County of Rockland* litigation (hereafter, Administrative Record).

[115] Section 102(2)(C) of NEPA requires all federal agencies to prepare an EIS for "major Federal actions significantly affecting the quality of the human environment . . .." 42 U.S.C. §§ 4332(2)(C). CEQ's NEPA regulations are set forth at 40 C.F.R. Part 1500 et seq.; FAA's NEPA requirements and guidance are set forth in FAA Order 1050.1E and in the *FAA Airspace Redesign Handbook.*

[116] 40 C.F.R. § 1502.13.

[117] 42 U.S.C. § 4332(2)(C)(iii); 40 C.F.R. § 1502.14.

[118] 42 U.S.C. § 4332(2)(C)(ii); 40 C.F.R. §§ 1502.16, 1508.7, 1508.8.

[119] 40 C.F.R. § 1506.6.

[120] Executive Order 12898, Sec. 1-101.

[121] Agency compliance with NEPA is evaluated according to § 10 of the Administrative Procedure Act (APA), 5 U.S.C. § 706, which provides that a court may "hold unlawful and set aside agency action, findings, and conclusions found to be arbitrary, capricious, an abuse of discretion or otherwise not in accordance with the law." This standard grants deference to agency decisions, with courts presuming that an agency's action is valid. *See Vermont Yankee Nuclear Power Corp. v. NRDC, 435 U.S. 519 (1978); Citizens to Preserve Overton Park v. Volpe*, 401 U.S. 402 (1971); *City of Olmsted Falls v. FAA*, 292 F.3d 261 (D.C. Cir. 2002); *North Slope Borough v. Andrus*, 642 F.2d 589 (D.C. Cir. 1980). To determine whether an agency's action was "arbitrary or capricious," courts assess whether the agency came to grips with the obvious ramifications of its approach and addressed them in a reasoned fashion. See NRDC v. EPA, 859 F.2d 156 (D.C. Cir. 1988). This requires an examination of the relevant data and an articulation of a satisfactory explanation for its action—including "a rational connection between the facts found and the decision made." Id. at 209 (citations omitted).
In the NEPA context, this "arbitrary and capricious" standard has been interpreted as requiring an agency act reasonably in carrying out NEPA's requirements (applying a "rule of reason") or that an agency take a "hard look" at the environmental consequences of its proposed actions. *See, e.g., Citizens Against Burlington*, 938 F.2d at 196 (upholding FAA approval of airport expansion); *City of Grapevine v. DOT*, 17 F.3d 1502, 1504 (D.C. Cir. 1994), cert. denied, 513 U.S. 1043 (1994) (same). As long as an agency meets this standard, a court should not substitute its own judgment, and the agency's action should be affirmed. *Kleppe v. Sierra Club*, 427 U.S. 390, 410 n.21 (1976) (upholding the Department of the Interior's decision not to conduct comprehensive EIS on speculative proposals). In addition to its basis in the APA, this deference stems from NEPA's imposition of procedural rather than substantive requirements—NEPA's purpose is to ensure that government agencies consider environmental consequences and act on full information, and that interested groups have access to such information, but NEPA does not mandate a particular environmental result. *Sierra Club v. United States Forest Service*, 46 F.3d 835, 837 n.2 (8th Cir. 1995). "[O]nce an agency has made a decision subject to NEPA's procedural requirements, the only role for a court is to insure that the agency has considered the environmental consequences . . ." *Strycker's Bay Neighborhood Council, Inc. v. Karlen*, 444 U.S. 223, 227-28 (1980).

[122] *See* Administrative Record, Document No. 9358, MITRE Report, "Eastern U.S. Problem Assessment and Multi-Cen." Among other things, MITRE modeled and simulated all alternatives for the airspace redesign project.

[123] Draft EIS, p. 2-2; final EIS, p. 2-2.

[124] A TRACON is an air traffic control facility that uses radar and two-way radio communication to separate air traffic within a specific geographic area in the vicinity of one or more large airports.

[125] Final EIS pp. 1-25.

[126] Final EIS pp. 1-21. In 49 U.S.C. § 40101(d), Congress declared national policy to be that FAA shall: "(1) assign[], maintain[], and enhanc[e] safety and security [as] the *highest* priorities in air commerce; . . . [and] (4) control[] the use of the navigable airspace and regulat[e] civil and military operations in that airspace in the interest of the safety and efficiency of both of those operations . . .." (Emphasis added.)

[127] *See http://www.faa.gov/about/mission/* (accessed on May 5, 2008). FAA has certain general noise reduction authority, see 14 C.F.R. Part 150, but noise reduction is not part of FAA's statutory mission.

[128] *See* S. Rep. No. 108-146, at 22 (2003), accompanying S. 1589 ("As FAA moves forward with its redesign program in the New York/New Jersey and Philadelphia area, the Committee encourages FAA, where appropriate, to consider air noise impacts as part of the redesign effort."). See also S. Rep. No. 107-224, at 43 (2002), accompanying S. 2808. Additionally, certain appropriations committee reports stated that FAA may not use funds for its New York/New Jersey/Philadelphia redesign NEPA review "as long as the FAA fails to consider noise mitigation," see H.R. Rep. No. 108-792, at 1366 (2004), accompanying H.R. 4818. See also H.R. Rep. No. 109-307, at 136 (2005), accompanying H.R. 3058; H.R. Rep. No. 109-307, at 136 (2005), accompanying H.R. 3058. Cf. H.R. Rep. No. 108-671, at 16 (2004), accompanying H.R. 5025 (stating FAA may not use funds for NEPA review of New York/New Jersey/Philadelphia redesign "unless the FAA formally declares noise mitigation to be a primary objective of the redesign project.") (statement not included in conference report).

[129] Final EIS, pp. 1-25, 1-26.

[130] Final EIS, p. ES-2.

[131] Noise Mitigation Report, FAA, Apr. 6, 2007.

[132] Final EIS, pp. 1-26.

[133] Final EIS, pp. 2-8.

[134] The experts we interviewed also found that FAA's purpose and need statement was reasonable and that the agency did not need to include noise or environmental concerns in its objectives.

[135] See footnote 16 above.

[136] Final EIS, p. ES-2.

[137] Final EIS, p. 2-8.

[138] Final EIS, p. 2-9.
[139] Final EIS, p. 2-9.
[140] These objectives included reducing airspace congestion, reducing environmental impacts, where possible, and accommodating projected growth. Final EIS, pp. 2-9 – 2-11.
[141] The alternatives considered in detail were the Future No Action Airspace Alternative (required by NEPA), the Modifications to Existing Airspace Alternative, the Ocean Routing Airspace Alternative, and the Integrated Airspace Alternative Variation with, and without, Integrated Control Complex (ICC) (ICC refers to a common automation platform). Final EIS, p. ES-4. FAA considered the Ocean Routing Alternative, even though it did not meet the purpose of increasing the efficiency and reliability of the New York/New Jersey/Philadelphia airspace, because of long-standing concerns of New Jersey Citizens Against Aircraft Noise, a citizens group. Final EIS, pp. 2-10, 2-11.
[142] App. IV to this report describes the selected project in more detail.
[143] Operational viability refers to whether a particular airspace redesign is workable and thus safe; criteria include reduced airspace complexity and reduced voice communications. Operational efficiency refers to how well a particular design works; criteria include reducing delay, balancing controller workload, meeting system demand, improving user access to the system, expediting arrivals and departures, increasing flexibility in routing, and maintaining airport throughput (throughput is explained in note 36 below).
[144] Final EIS, Chapter 4.
[145] 40 C.F.R. § 1502.14(d).
[146] 40 C.F.R. § 1502.14(c).
[147] Although we conclude FAA's selection and analysis of alternatives met the legal requirements of NEPA, we nevertheless have identified deficiencies in FAA's analysis. See pp. 36-45 of this report.
[148] Final EIS, Appendix N, p. 1164. FAA further explained: "Throughput is not the same as capacity. Throughput is the actually achieved number of aircraft using a resource in a given time. It is measured by counting aircraft, whether in a real system or a simulated one. Capacity is the theoretical maximum number of aircraft that could use a resource in a given time. It is measured by surveying, queuing simulations, or mathematical models. A decrease in throughput does not mean a reduction in the number of flights; it means that delays increase. Likewise, an increase in throughput does not mean an increase in flights, it means a decrease in delays. When throughput is below capacity, the system is inefficient. Reducing the difference between the throughput and the capacity is the purpose of this airspace redesign . . .." *Id.*
[149] *See* Administrative Record Document No. 8239, Meeting minutes with NY DOT, Dec. 17, 2003 ("airspace redesign is not related to increasing capacity because the capacity of the system is currently limited by the pavement (runways) at the airports. Airspace redesign produces a more efficient system and reduces delay.").
[150] While we believe FAA's decision not to assess environmental impacts from possible growth from the redesign would likely survive challenge as a legal matter, particularly given the deferential standard of review, we nevertheless find FAA's induced-growth analysis contained several limitations, as discussed at pp. 36-40 of this report.
[151] 40 C.F.R. § 1501.7.
[152] 40 C.F.R. § 1506.6(a).
[153] 40 C.F.R. § 1506.6(b).
[154] 40 C.F.R. § 1506.6(c).
[155] 40 C.F.R. §§ 1503.1(4), 1506.6(d).
[156] 40 C.F.R. § 1503.4(a).
[157] FAA Order 1050.1E, § 208b.
[158] FAA Order 1050.1E, §§ 208d, 508(g).
[159] FAA Order 1050.1E, § 208c.
[160] FAA Order 1050.1E, § 505a.
[161] FAA Order 1050.1E, § 208c.
[162] Final EIS Table 6.1; Record of Decision, pp. 28, 48.
[163] A summary of the public comments and FAA's responses to these comments are contained in appendices N and Q of the final EIS.
[164] Additional detail on the public comment periods is contained in table 1, p. 26.
[165] See final EIS, appendices N and Q.
[166] DOT Order 5610.2; 62 Fed. Reg. 18377 (Apr. 15, 1997).
[167] CEQ Guidance, p. 5.
[168] CEQ Guidance, p. 8.
[169] CEQ Guidance, p. 10.
[170] CEQ Guidance, p. 9.
[171] Presidential Memorandum, p. 1.
[172] CEQ Guidance, p. 9.
[173] CEQ Guidance, pp. 10-11.

[174] CEQ Guidance, p. 11. Examples of how agencies should enhance their outreach include the use of religious organizations, newspapers, radio, and other media, civic associations, community and social service organizations, labor organizations, and the internet and other electronic media.

[175] CEQ Guidance p. 12.

[176] CEQ Guidance, p. 13.

[177] CEQ Guidance, p. 13.

[178] CEQ Guidance, p. 13.

[179] CEQ Guidance, p. 13.

[180] CEQ Guidance, p. 13.

[181] CEQ Guidance, pp. 15-16.

[182] FAA first identified the census blocks significantly impacted by aircraft-related noise and determined the minority composition and median income of those census blocks based on prospective census data for 2011. *See* final EIS, tables 3.8, 3.9, showing minority populations in the study area and poverty population statistics. To determine whether there would be a high and adverse impact to a low-income community, the median income of the significantly impacted census blocks was compared to the poverty-level income. Additionally, FAA used CEQ's definition of a minority population. FAA then determined whether a low-income or minority community subjected to a high and adverse human health or environmental impact was disproportionately affected, deemed to occur when an environmental impact is either predominantly borne by a low-income or minority population or is more severe in magnitude than the impact on the non-low-income or nonminority populations. *See* final EIS, pp. 4-43, 4-44.

[183] *See* final EIS, Chapter 4 and Appendix I.

In: FAA Airspace Redesign and Congestion Management ISBN: 978-1-60692-710-6
Editor: Sidney R. Rothman

Chapter 2

AVIATION CONGESTION AND DELAY: SYSTEM-WIDE AND NEW YORK-AREA ISSUES

Bart Elias, John W. Fischer and Todd B. Tatelman
Resources, Science, and Industry Division,
American Law Division

SUMMARY

Aviation congestion and delay is an issue of long standing. Statistically, there was a significant increase in congestion and delay throughout the national aviation system in 2007. The situation has been especially noticeable at certain key airports, namely the New York region in general and John F. Kennedy International Airport (JFK) in particular. The Department of Transportation (DOT) and its operating agency, the Federal Aviation Administration (FAA), have promised to take actions aimed at reducing congestion and delay both in the short and long terms. In the short term, DOT is proposing to address the JFK situation through administrative and economic measures that would likely restrict or otherwise provide for the allocation of flights into the airport during specific periods of time.

In addition to the proposals likely to be made by DOT, there are other potential near-term fixes that can be considered to add system-wide capacity to the aviation system. These include, but are not limited to: airspace redesign, use of military airspace, and airspace flow control. Over the longer term the FAA is primarily relying on modernization of the air traffic control system through the Next Generation Air Transportation System (NGATS or NextGen) and the creation of new airport infrastructure to provide major reductions in aviation congestion and delay, although a recent FAA capacity needs study has concluded that these enhancements are needed immediately, especially in the New York region.

Congress has taken an interest in this issue as growing concerns over airline service have paralleled ongoing legislative action to reauthorize the activities of the FAA and the revenue mechanisms for funding the operations of the national airspace system. The Bush Administration, and especially Secretary of Transportation Peters, favor examining market mechanisms as a means to pay for infrastructure operations, maintenance, and development in all transportation modes. When the Administration submitted its proposals for FAA reauthorization in early 2007 (H.R. 1356/S. 1076), it included a pilot program to evaluate market-based mechanisms to relieve congestion at up to 16 airports. The FAA proposal was not adopted in either the House (H.R. 2881) or Senate (S. 1300) reauthorization bills still under consideration at the end of 2007, but the House bill contains provisions intended to help the FAA deal with congestion problems at specific airports.

Regulation of the national air transportation system is legally the domain of the federal government. The operation and ownership of airports, however, is provided primarily by regional, state and local entities. Neither the FAA nor the DOT have, to date, attempted to develop and impose a congestion pricing scheme on the local authorities. In the event that FAA should decide to make such an attempt, it appears that there are several legal issues that may arise.

Aviation Congestion and Delay: System-Wide and New York-Area Issues

Aviation congestion and delay is an issue of long standing. Statistically, there was a significant increase in congestion and delay throughout the national aviation system in 2007. The situation has been especially noticeable at certain key airports, namely the New York region in general and John F. Kennedy International Airport (JFK) in particular. The Department of Transportation (DOT) and its operating agency, the Federal Aviation Administration (FAA), have promised to take actions aimed at reducing congestion and delay both in the short and long terms. In the short term, DOT is proposing to address the JFK situation through administrative and economic measures that would likely restrict or otherwise provide for the allocation of flights into the airport during specific periods of time. More detail on DOT's plans is expected to be available in the latter half of December 2007.[1]

In addition to the proposals likely to be made by DOT, there are other potential near-term fixes that can be considered to add system-wide capacity to the aviation system. These include, but are not limited to: airspace redesign, use of military airspace, and airspace flow control. Over the longer term the FAA is primarily relying on modernization of the air traffic control system through the Next Generation Air Transportation System (NGATS or NextGen) and the creation of new airport infrastructure to provide major reductions in aviation congestion and delay.

This chapter discusses the causes of congestion and delay in the current context. It also discusses some technological and administrative mechanisms that might be employed to mitigate these problems. This discussion is followed by an overview of demand management (economic) remedies to congestion as they might be applied in an airport/airspace environment, and discusses the rather limited historical experience with the use of these mechanisms. Finally, the chapter concludes with a legal overview of how specific congestion remedies, in particular those involving aviation system pricing, might be applied and administered.

AIR CARRIER DELAY TRENDS

Over the past four years, system-wide airline passenger volumes have continued to grow, now surpassing pre-9/11 levels. Along with this growth, flight delays and cancellations have also increased, and statistics indicate that the percentage of flights delayed 15 minutes or more in 2007 may surpass the prior record set in 2000. The percentage of cancelled flights is also up, although not quite as high as the percentage of cancellations experienced in 1999, 2000, and 2001 (see **Figure 1**). Some also point out that delay statistics alone do not tell the whole story. This is because airlines have, in many cases, opted to lengthen scheduled flight times, particularly along heavily congested and frequently delayed routes.[2] A comparison of present day airline schedules to 1997 schedules found that published trip times are about 10% higher than they were 10 years ago.[3] Additional schedule adjustments could be forthcoming and may further increase typical travel times, particularly in response to recent Administration announcements that it would consider operation of chronically delayed flights to constitute an "unfair and deceptive practice."[4]

Despite airline schedule adjustments that have increased planned-for trip times, airline delays are still ticking upward. As a consequence of this increase in delays, combined with airline schedule padding and passengers factoring additional wait times at security screening checkpoints into their travel plans, airline passenger trips are getting longer on average and travelers often face more hassles. Consumer complaints and media coverage of air carrier service deficiencies and cases of extensively delayed flights have put the specific issue of air carrier flight delays in the spotlight during the summer of 2007. In response, the DOT has convened a task force consisting of senior management to assess persisting problems with airline delays and consumer service, and has convened a special aviation rulemaking committee (ARC), consisting of aviation stakeholders representing airlines, airports, general aviation users, and airline consumers, to specifically address delays and other airline service issues in the New York metropolitan area. Based on the findings and recommendations of these groups, which are expected before the end of calendar year 2007, the DOT intends to propose options to address and remedy airline delay and customer service problems.[5] Congress has taken a particular interest in this issue as these growing concerns over airline service have paralleled ongoing legislative action to reauthorize the activities of the FAA and the revenue mechanisms for funding the operations of the national airspace system.

While there are broad concerns over aviation congestion and delay at many of the nation's busy airports, there is particular concern regarding airline service and performance at major airports in the New York metropolitan area where delay problems are the most acute. On-time arrival and departure statistics for the past two years compiled by the DOT's, Bureau of Transportation Statistics (BTS) have consistently ranked the three New York metropolitan area commercial passenger airports — Newark Liberty (EWR), John F. Kennedy International (JFK), and LaGuardia (LGA) — at or near the bottom of the list among the busiest 32 airports in the United States. From January 2007 through August 2007, about one out of every three flights arrived or departed more than fifteen minutes late at these airports.

Between January 2002 and September 2007 at these three airports, arrival and departure delays were typically around one hour in length. At Newark Liberty Airport, where delays have consistently been the worst among busy U.S. airports, departure delays averaged 67 minutes and arrivals delays averaged 73 minutes over the time frame from September 2006 through September 2007.[6] According to the FAA, over the summer of 2007, 44% of all air delays experienced throughout the United States occurred in the New York, New Jersey, and Philadelphia region.[7] What's more, there was a dramatic increase in the percentage of flights cancelled at all three of these airports in 2007 compared to prior years. From January through September 2007, JFK saw 3.18% of its flights cancelled, LGA had 4.81% of its flights cancelled, and at EWR, 3.65% of scheduled passenger flights were cancelled. Not only are these cancellation rates much higher than previous years, but they are also considerably higher than the system-wide averages presented in Figure 1.

Causes of Air Carrier Delays

While many travelers perceive that delays are frequently associated with weather, actual delays directly attributable to severe weather conditions account for only a relatively small portion of total system-wide delays (about 6% of all air carrier delays from June 2003 through August 2007). Rather, delays are most readily attributable to a combination of the current system's inability to cope with weather, congestion, and other factors affecting the efficient flow of traffic at major airports and along crowded airways. These types of delays, referred to as national aviation system delays, account for about 31% of all air carrier delays and are the types of delays that the DOT believes can be remedied through investment in new technologies and more efficient procedures for air traffic management. Other sources of delays include air carrier delays arising from maintenance difficulties and inefficiencies in air carrier operations which account for about one quarter (27%, to be exact) of all delays.

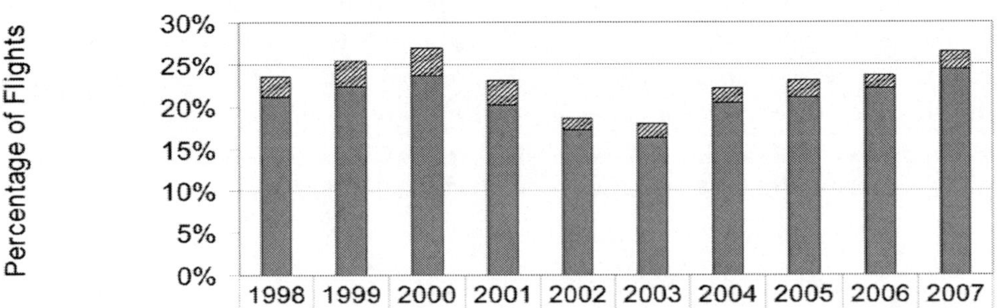

	1998	1999	2000	2001	2002	2003	2004	2005	2006	2007
Cancelled (%)	2.44	3.10	3.36	2.78	1.29	1.61	1.66	1.97	1.53	2.24
Delayed (%)	21.13	22.33	23.68	20.25	17.27	16.28	20.46	21.06	22.09	24.33

Calendar Year

Delayed (%) Cancelled (%)

Figure 1. Annual System-Wide Air Carrier Flight Delays and Cancellations (1998-2007)
Source: Bureau of Transportation Statistics.
Note: Statistics for 2007 cover 1/07 through 9/07.

Also, the cascading effects resulting from aircraft arriving late that cannot be turned around in time to maintain outbound flight schedules account for 36% of all flight delays. However, the root causes of initial schedule delays that produce these cascading delays throughout an operational day can be primarily traced back to system inefficiencies and air carrier scheduling inefficiencies and constraints. Despite heightened aviation security, security related delays have had a negligible impact (less than 1% of flight delays) on air carrier operations (see Figure 2).

Analysis of airline schedule delays and cancellations across the entire system, and more specifically at the three New York area airports, point to no one single cause, but rather a variety of potential contributing factors. These factors can be grouped into three broad categories:

- *Airline Scheduling and Equipage*
- *Constraints on Airport and Air Traffic Control System Capacity; and*
- *Business Jets and Other High-Performance General Aviation Aircraft Operations*

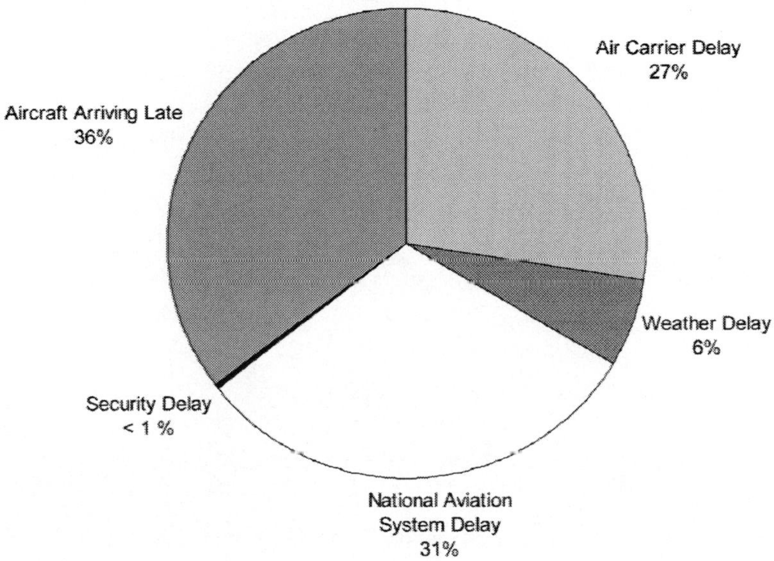

Figure 2. Causes of Air Carrier Flight Delays (June 2003-September 2007)
Source: CRS Analysis of Bureau of Transportation Statistics Delay Cause Summary Data.

Airline Scheduling and Equipage

The single leading cause of delays is the cascading effects of aircraft arriving late. Examining the trends in flight delay causes since June 2003 reveals that the largest increase among delay factors is also late aircraft arriving. This may, in part, be an indicator of possible airline over-scheduling and/or increased fleet utilization. In recent years, efforts to cut operational cost at airlines have driven airlines to reduce turn times (i.e., the time aircraft spend parked at the gate being serviced between flights). These data suggest that either the airlines themselves, the airport infrastructure, the FAA air traffic management system, or some combination of these aviation system components cannot fully accommodate this increased operational tempo. The result is cascading delays as an operational day progresses. This is also evidenced in statistics showing an increase in delays as it gets later in the operational day (see Figure 3).

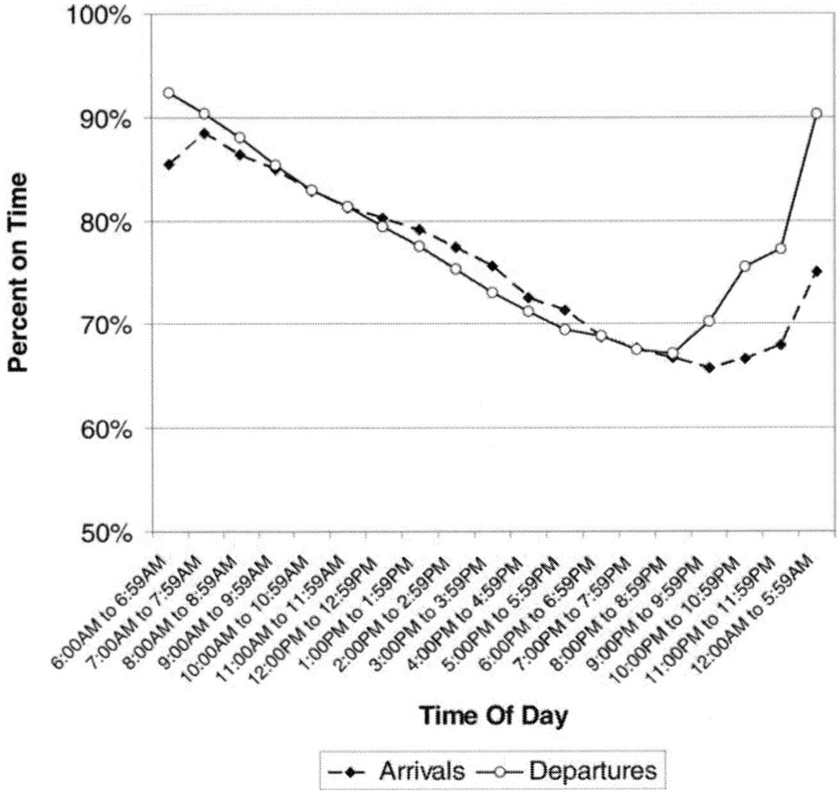

Figure 3. System-Wide Percent of Flights Arriving and Departing On Time as a Function of the Time of Day (2006 Data)
Source: CRS Analysis of Bureau of Transportation Statistics Airline On-Time Performance Data.

It seems somewhat paradoxical that the increased operations tempo can actually contribute to increased delays. To put this into perspective, it is useful to examine air carrier fleet trends. In 2000, mainline passenger airlines operated 4,488 jets compared to an estimated 3,886 in 2006. Over the same period, the number of aircraft operated by regional carriers, which are primarily smaller regional jets, grew from 2,274 in 2000 to an estimated 2,743 in 2006. Taking these two industry components together, the number of operating passenger aircraft declined from 6,762 in 2000 to an estimated 6,629 in 2006, roughly a 2% decrease. Looking solely at large jets, however, the trend is much more evident. Large jets in service have declined from 4,462 in 2000 to an estimated 3,827 in 2006, roughly a 14% decline. Whereas large jets made up about 66% of the total scheduled passenger service fleet in 2000, today large jets make up about 58%. In terms of the percentage of jet aircraft in airline operations, large jets made up about 88% of all jets used in scheduled passenger service in 2000, whereas in 2006, large jets comprised only 68% of all operating scheduled passenger jets. The trend over the past five years has clearly been toward reducing the number of large jets and increasing the number of smaller regional jets.[8] So, not only are there fewer aircraft in service compared to 2000, but the aircraft that are being utilized are, on average, smaller in size and hold fewer passengers. As passenger demand for air travel continues to grow, these two trends can contribute to increased congestion and the need to sustain increased operations tempo to handle this growing passenger demand. With an increased operations tempo and utilization of smaller aircraft, there is greater demand for airport gates, so delays to one flight — say for weather or maintenance-related issues — can have cascading effects on other flights because of a lack of gate availability.

The shift toward using smaller aircraft has been a major contributor to the increased traffic at the three New York metropolitan area airports. For example, a comparison of August 2002 airline schedules to August 2007 schedules found that, over the past five years, there has been a 128% increase in the use of aircraft with fewer than 100 seats and an increase of 120% in flights using aircraft with 100 to 200 seats at JFK. Over that same time period, flights at JFK utilizing aircraft with more than 200 seats declined by 12%. This illustrates a dramatic shift in the type and scale of operations at JFK, which has long been regarded as a key origin and destination for international flight operations with widebody aircraft historically making up a sizable portion of the flight activity at the airport. The situation has been similar but somewhat less pronounced at LGA and EWR. Over the past five years, LGA experienced a 49% jump in scheduled flights of aircraft with fewer than 100 seats, but about a 8% decline in operations using aircraft with more than 100 seats. At EWR the shift hasn't been quite as dramatic, but is nonetheless still considerable. Over the past five years, utilization of aircraft with fewer than 100 seats at EWR has increased by almost 34%, while there has been relatively little change in the number of flights using aircraft with more than 100 seats, based on the comparison of August 2002 and August 2007 flight schedules.[9]

In addition to this increased operations tempo using smaller aircraft, a variety of other factors may influence airline scheduling practices, some of which may run counter to initiatives for reducing congestion and delay. For example, current LGA slot rules impose "use it or lose it" conditions on the airlines. Airlines have responded by continuing to operate unprofitable flights rather than concede these coveted slots to their competitors. For example, in June 2007, it was reported that there were three unprofitable markets, comprising a total of 13 daily flights, operating at LGA.[10] While utilizing large aircraft with high load factors (high percentages of filled seats) would be best for optimizing limited available capacity at congested airports like LGA, reviews of airline scheduling practices indicate that continued use of smaller aircraft operating routes which generate low load factors leads to suboptimal utilization in scheduling. Rationales for such strategies among the airlines include offering consumers greater flexibility in scheduling and serving smaller communities in order to capture or control greater market share at capacity constrained airports. There are also political considerations and incentives in place to maintain service to smaller communities, for example through Essential Air Service (EAS) subsidies. Due to intense competition for market share at high density airports, including the New York area airports, airlines have no particular incentives to reduce or otherwise modify flight schedules. While chronic delays and customer dissatisfaction would appear to provide some impetus for airlines to take action, each airline individually faces a "prisoner's dilemma," because they fear that if they reduce schedules or shift flights to off-peak periods, their competitors could gain an advantage by filling in any schedule openings this might create.[11]

Constraints on Airport and Air Traffic Control System Capacity

The aviation system in the United States suffers from the fact that flight operations are highly concentrated at a relatively small number of airports. The 35 busiest airports in the United States handle about 75 percent of all airline passengers. This is largely because demand for air travel is highly concentrated within the country's major metropolitan regions. The New York area, being the most highly populated metropolitan region, generates a demand for air travel that significantly strains available capacity. This strain on capacity is not limited to commercial airline travel, as it also extends to demand for access to airports and airspace by corporate and private jets and fractionally-owned and chartered aircraft. The major New York area airports, as well as general aviation reliever airports in the area, have limited options for expansion. An updated FAA/MITRE Corporation study of capacity needs in the national airspace system released in May 2007 concluded that "[a]dditional capacity within the New York Metropolitan Area is needed."[12] While this conclusion may now seem obvious even to casual observers, the FAA was slow to reach this conclusion and the original capacity needs study released in 2004 did not identify the New York metropolitan area as immediately needing additional capacity.

Many regard Stewart International Airport (SWF), a converted military airbase located about 55 miles north of Manhattan, as potentially providing needed capacity to this region, although the FAA's capacity needs study concedes that "it is too early to determine whether [NY/NJ Port Authority acquisition of SWF] will offset the need for additional capacity enhancements."[13] Distance from Manhattan and lack of ground transportation infrastructure are two potential challenges for making effective use of SWF to relieve congestion at the three major New York area airports, yet many remain hopeful since options to otherwise expand capacity in the region appear quite limited.

While the focus of current media attention and policy debate has been on congestion and delay in the New York metropolitan area, this region is not alone in its need for additional airport and airspace capacity. The FAA/MITRE Corporation 2007 capacity needs study identified 14 airports and eight metropolitan regions that will need additional capacity beyond that anticipated from ongoing and planned airport expansion projects over the next 17 years.[14] Since further airport expansion is a limited option, experts are focusing heavily on looking at ways to overcome technical and procedural constraints on airspace design and utilization in congested areas.

Airspace in congested metropolitan areas is largely constrained by air traffic controller workload and safety considerations. Emerging technologies to improve the accuracy of aircraft navigation and surveillance by primarily relying on precision satellite-based navigation capabilities, thereby allowing more efficient utilization of airspace and greater flexibility in routing flights, are a key component of the Next Generation Air Transportation System (NGATS) Concept of Operations (CONOPS).[15] However, even if systems development and acquisition is kept on schedule, this future air traffic control system will not be fully operational for another 17 years. In the meantime, the FAA has sought to improve the efficiency of the current air traffic system through airspace redesign initiatives and procedural changes using currently available technologies. For example, in 2001 the FAA implemented its "National Choke Point Initiative" to reduce airspace bottlenecks at several points in the Eastern United States, including in the airspace serving the New York metropolitan airports. This initiative included procedural changes and adding new air traffic control sectors to better distribute controller workload. It was reported that these initiatives had the effect of reducing delays by 20%.[16] Also, the recently approved NY/NJ/PHL Airspace Redesign is an initiative to specifically restructure airspace and air traffic flows in the New York and Philadelphia regions to improve efficiency and reduce delays. According to the FAA, this redesign initiative could also reduce delays by 20%.[17] These initiatives have been combined with procedural measures, such as ground delays and airspace flow programs as well as the implementation of various decision-aiding tools to assist in air traffic management. These approaches are largely regarded as tactical stopgap measures to mitigate delay conditions, but are not seen by themselves as complete long term solutions for reducing delays as traffic volumes continue to increase.

Despite some success in reducing delays through these initiatives, traffic growth is having the effect of offsetting many of the gains achieved by taking these steps to improve system efficiency. FAA airspace initiatives in the New York area appear to be having a difficult time keeping up with the growth in demand for air travel in this region. Aviation experts largely agree that a comprehensive systems-based approach to implementing new technologies and flight procedures is needed to most effectively address capacity needs and reduce delays across the national airspace system, particularly in the New York metropolitan region airspace, where chronic delays are symptomatic of constrained airport and airspace capacity.

The airline industry in particular is advocating for the FAA to take steps to maximize the utilization of airport and airspace capacity in the New York metropolitan region, arguing that demand-management strategies are, in their opinion, not necessary.[18] The Air Transport Association (ATA) has specifically called for establishing a New York area air traffic delay management "czar" to coordinate delay-reduction initiatives, and involving air carriers and other airspace users in discussions to identify near-term operational and procedural measures to address system inefficiencies. While the work of the special ARC announced by the DOT would appear to serve this purpose, it remains unclear what specific solutions this committee may be able to offer that can improve system efficiency beyond what has already been tried and implemented under the ongoing NY/NJ/PHL airspace redesign and other airspace and operational initiatives, such as ground delay and airspace flow management programs that are currently in place. Any further initiatives not yet implemented may only be able to offer smaller marginal improvements toward increasing system efficiencies and reducing delays. Therefore, despite opposition from the airline industry, many academic experts believe that a variety of demand-management or market-based approaches for addressing congestion and delay at various high-density or capacity constrained airports, including the three commercial passenger airports serving the New York metropolitan area may need to be considered in the near term.[19]

Business Jets and Other High-Performance General Aviation Aircraft Operations

The role that business jets and other high-performance general aviation aircraft play in air traffic system congestion and air carrier delays remains both debatable and highly contentious. At busy air carrier airports, general aviation makes up a relatively small percentage of flight operations. According to the FAA FY2005 Cost Allocation Report, general aviation aircraft and fractionally-owned aircraft accounted for less than 4% of all aircraft operations at large hub airports in FY2005.[20] In FY2005, general aviation made up 2.7% of the traffic at LGA, 3.2% of the traffic at EWR, and 2.0% of the traffic at JFK.[21]

The airlines, however, have argued that general aviation operations in terminal airspace and along the busiest en route corridors place a considerable strain on system capacity, pointing out that the number of private jet aircraft operating in U.S. airspace has grown from about 1,800 in 1970 to almost 18,000 today.[22] The airlines argue that "[t]he proliferation of these aircraft, in addition to the unprecedented growth of the business aviation sector, has been the primary factor in the increased workload for FAA."[23] In the New York area, there is some evidence of the additional demand for air traffic services from general aviation. At Teterboro Airport (TEB) in Teterboro, NJ — a major GA reliever for the New York area — total operations counts have remained flat over the thirty years from 1976 to 2006 (the data actually show a slight decline of 2.3% in total operations for TEB). Over the same period, however, the number of instrument operations have increased threefold (306.3% increase in 2006 compared to 1976 levels). Whether and to what extent this increased general aviation utilization of air traffic services is affecting airline operations and contributing to delays at nearby airports has not been conclusively determined, but it remains a specific point of contention among the airlines and general aviation interest groups such as the National Business Aviation Association (NBAA). Since high-performance general aviation aircraft typically do not make up a large percentage of the air traffic flying directly to and from capacity constrained commercial airports, their impact on flight delays could most likely be addressed through airspace redesign and new technologies and procedures for air traffic management.

Some Possible Near-Term Remedies for Reducing Air Traffic Congestion and Mitigating Delays

A variety of options for reducing air traffic congestion and minimizing airline delays are under consideration. While most aviation industry experts believe that the largest gains can be achieved in the long term through investment in Next Generation Air Transportation System (NGATS or NextGen) technologies, the current situation is prompting many in government and the aviation industry to seek out near-term fixes to address congestion and delay, especially in the New York area. A variety of near-term options are under consideration including phasing-in airspace redesign plans in the New York/New Jersey/Philadelphia region; opening military airspace to civilian flights under certain conditions; implementing ground delay programs, airspace flow control programs, and other congestion management tools to improve the flow of air traffic; phasing-in technology and procedural changes to reduce aircraft separation and improve airspace utilization; and possibly improving airline schedule optimization at high density airports by utilizing larger aircraft, a practice known in the industry as upgauging.

Airspace Redesign

As mentioned earlier, the FAA recently finalized its plans for redesigning airspace in the New York, New Jersey, and Philadelphia areas (the NY/NJ/PHL Airspace Redesign). In the near future, the FAA will begin implementing this plan which is touted to more efficiently use airspace and runways at the New York area airports and at Philadelphia International Airport (PHL). The plan includes greater use of fanned departure headings and parallel arrivals, as well as greater flexibility in traffic flows to handle severe weather conditions. The FAA anticipates that implementation of the NY/NJ/PHL Airspace Redesign will cut delays by a cumulative total of 200,000 hours annually, resulting in a savings of $248 million annually in airline operating costs, plus another $37 million annually because of the increased flexibility during severe weather.[24] These redesign initiatives probably offer the greatest near-term potential for reducing delays at the New York area airports without curtailing or otherwise modifying airline schedules.

Opening of Military Airspace to Civilian Air Traffic

Currently, the FAA negotiates in real-time with the DOD to open up airspace to civilian traffic on an as-needed basis, such as to detour flights around thunderstorms.[25] However, prior to the 2007 Thanksgiving holiday travel period, President Bush announced a plan to open military airspace along the east coast to airline flights during a five day period from the day before Thanksgiving through the Sunday after Thanksgiving. During this time, the FAA established two additional east coast flight paths, dubbed the "Thanksgiving express lane" by Administration officials.[26] This prearranged opening of military airspace over the entire Thanksgiving weekend was touted by some as an "unprecedented step"[27] while dismissed by others as giving "[air traffic] controllers a power they already have and have used for years,"[28] namely the authority to utilize military airspace during holiday periods when the military is typically not training. The Air Transport Association (ATA) described the utilization of military airspace over this period as being moderate, ranging between 10 to 40 aircraft per hour being routed along these temporary flight paths.[29] Whether the use of this airspace helped reduce or mitigate delays is hard to say, particularly at the New York area airports where delays over the 2007 Thanksgiving holiday were largely attributed to low ceilings and reduced visibility which impacted arrival and departure spacing. In the past, use of military airspace has been on more of a case-by-case basis, such as to reroute traffic around weather, leading some to consider whether more extensive use of military airspace along the east coast could be a possible fix for airspace congestion and resulting delays.

Experts have mixed opinions regarding the degree to which more extensive use of military airspace can relieve airspace congestion and mitigate delays, especially if military airspace is opened only for sporadic, temporary intervals. This is because the FAA's airway structure and arrival and departure routes along the east coast have been designed in coordination with the DOD taking into consideration those areas carved out for military use.

Turning on and off civilian flight activity through these areas would require continued close coordination between the FAA and the DOD, and this is not something that the relatively rigid current airspace structure and procedures were specifically designed for. Military airspace, however, is often inactive (that is, not utilized for any military purpose), thus offering potential options for more direct routing of civilian flights and additional flight paths to alleviate airspace congestion. The difficulty is knowing when military airspace will be inactive and could be made available for civilian use. Better coordination of airspace scheduling between the DOD and the FAA could potentially open additional routes and give air traffic controllers greater flexibility in routing flights during certain periods of the day, particularly as more and more aircraft are allowed to utilize direct routing capabilities relying on satellite-based navigation and are thus not as dependent on the rigid legacy airway structure. Nonetheless, most concede that the overall impact of such actions would likely be relatively small, although perhaps not insignificant, because they do not address the primary issue of airport capacity constraints at the busiest airports along the east coast.

Ground Delay and Airspace Flow Programs and Other Congestion Management Tools

A variety of decision aiding tools to help airspace managers and air traffic controllers better optimize the utilization of airspace and minimize the disruption caused by significant weather and other events have been developed and are being continually refined. One approach has been to implement ground delays at origin airports for specific destination airports, like LGA, to better meter the flow of arrivals, thereby reducing arrival delays and the additional congestion of already crowded airspace that this may cause. In addition, airspace flow programs (AFPs) have been used to limit congestion in specific airspace sectors. Ground delays and AFPs can also be imposed on general aviation aircraft to limit their disruption of traffic flows to and from major airports that could trigger delays. Other decisionaiding tools that the FAA has developed for improving efficient utilization of airspace include the flight schedule monitor (FSM), for managing ground delays and airspace flow implementation, and the flight schedule analyzer (FSA), used to assess the effectiveness of ground delay and airspace flow procedures thereby allowing incremental improvements. Also the traffic management advisor (TMA) aids controllers in optimizing arrival sequences into major airports, and an adaptive compression program can aid in identifying controllers and reallocating slots that might otherwise go unused.[30] The FAA has also relied on collaborative decision making tools and operational "playbooks" for minimizing disruptions and delays caused by severe weather and other factors. Improving upon these various operational tools and techniques for optimizing air traffic flows and airspace utilization is an ongoing FAA initiative. However, there is concern that without accompanying technological advances in navigation and surveillance capabilities, these efforts will yield diminishing marginal improvements in delay reduction.

Technology and Procedural Changes

Airport traffic is constrained or "metered" at several points in the aviation system. On the airport itself, aircraft are constrained by the number of available gates, and arrivals and departures are metered by runway occupancy and arrival and departure spacing requirements. In the air, specific entry and exit points into and out of terminal airspace, referred to as gates, also constrict traffic flows in congested areas, and aircraft separation requirements impose limits on air traffic density. Technology to improve navigation and surveillance accuracy is seen as a long-term solution to safely increase traffic density and increase the flexibility of airspace utilization in congested terminal airspace and along busy routes. While the FAA has offered that the Next Generation Air Transportation System (NGATS or NextGen) provides a great hope for alleviating airspace capacity constraints and system delays, even if the system development stays on schedule, it will not, as mentioned earlier, be fully implemented for another 17 years.

By exploiting the improved navigational precision of aircraft satellite-based navigation systems using required navigation performance (RNP) and area navigation (RNAV) procedures, the FAA and airline operators can achieve better utilization of airspace, thereby increasing system capacity and potentially reducing delays. Examples include increased numbers of fanned departure headings to increase airspace utilization for departures, more closely spaced approaches to parallel runways, increased use of direct routing capabilities, and greater flexibility for routing aircraft around weather systems. These capabilities demonstrate how emerging satellite-based technologies have the potential to create opportunities for procedural changes that can significantly improve system efficiency, thereby potentially reducing delays. At the New York area airports and at PHL, RNAV/RNP arrivals and departures have recently been made available on a limited basis, and it is anticipated that they will be utilized to a much greater extent as implementation of the NY/NJ/PHL Airspace Redesign progresses.

Airline Upgauging

Another option for reducing congestion is to provide service using smaller numbers of larger aircraft. The selection of larger aircraft is referred to in the airline industry as upgauging. By upgauging aircraft in combination with schedule reductions, airlines could serve the same number of passengers with fewer flights. The airline trend in recent years, however, has been to "downgauge" rather than upgauge. This has been a particularly prevalent trend over the past five years as airlines have sought ways to increase load factor (the ratio of seats filled with revenue paying customers to total available seats) and reduce operating costs in a highly competitive market. At airports where there is significant competition, airlines have also sought to provide customers with the greatest amount of scheduling flexibility, offering more frequent flights, typically on smaller aircraft to keep load factors high and operations profitable. Without any particular incentives for upgauging, airlines are likely to be reluctant to do so on their own accord for fear that, by offering fewer flights to customers, they would be put at a competitive disadvantage. Airlines are also faced with the task of optimizing fleet composition in a manner that serves a broad array of markets in the most cost effective manner. So, while upgauging aircraft to serve highly competitive markets may help alleviate congestion and delay problems at specific airports, these larger aircraft may not be the best options for a particular airline's overall route structure. Determining an airlines optimum fleet mix is a complex task for airlines that serve a wide variety of markets, so simply upgauging aircraft serving capacity constrained airports may not necessarily be in the best interests of the airlines.

Demand Management: Slot Controls and Congestion Pricing

Demand management is a somewhat broad term that suggests that capacity, hence congestion, can be controlled using administrative or economic mechanisms, or a combination of the two. Historically, aviation usage of the term demand management is associated with airport slot controls and the allocation of slots. Over the last 4 decades slot controls have been imposed using both administrative and economic measures at 4 airports in the United States. During this period, as will be discussed below, there were attempts to create a market for slots, but these markets have never lived up to the expectations of their proponents.

Congestion pricing, the focus of several policy statements by Secretary of Transportation Mary Peters, can be viewed as a form of demand management that tries to use pricing techniques to effect desired results in reducing congestion.[31] Congestion pricing can be imposed at the airport level, for example, as part of, or as an adjunct to, landing fees. As will be discussed, there is some limited experience with the use of these mechanisms in the United States and in the United Kingdom (UK). Congestion pricing could also be applied to air traffic control (ATC) activities, though to date this mechanism has not been specifically applied in the United States or elsewhere.

The use of demand management techniques to ration finite amounts of transportation infrastructure has long been supported by economists. Americans, however, appear to have an aversion to tolls and the other transportation pricing mechanisms that are the stock and trade of demand management practice. Roads in this nation, for example, are provided largely as free goods, especially in the minds of drivers who believe they pay for roads sufficiently through taxes at the gas pump. This same framework also exists in the aviation world, with the airline ticket tax and associated taxes serving as a proxy for the highway fuels tax. Broadly based taxes, such as fuel taxes, however, are viewed by economists as proxies for user fees rather than as true user fees. As a result, taxes are generally seen as a relatively poor mechanism for pricing infrastructure. Nonetheless some observers, find it rather remarkable that a nation that differentiates access to many of its infrastructure systems on a price basis, e.g. telecommunications peak period pricing, has no comparable mechanism that differentiates for use of the national aviation system.

Congestion pricing is normally opposed by many groups in the aviation community. Historically, the groups most vociferous in their objections to this mechanism are regional airlines, business aircraft operators, and general aviation aircraft operators. All of these groups believe that congestion pricing could be used to discriminate against them and will have the effect of reserving what are publicly supplied assets — airports and ATC — for large airlines. It should be pointed out that the airline industry has also objected to certain demand management proposals and is objecting to the imposition of congestion pricing at JFK.[32] Residents of small and medium sized communities, primarily served by regional airlines, are also concerned that congestion pricing will price them out of major airports and further limit their access to the national aviation system.

Slot Controls

Well publicized capacity and delay problems in the late 1960s played an important role in the congressional debates that led to passage of the Airport and Airway Development and Revenue Acts of 1970 (P.L. 91-258, 1970 Act). The 1970 Act is seen as a milestone in that it created a programmatic framework for the modernization of the National ATC and airport systems, and provided a funding mechanism to pay for many of the needed upgrades — the airport and airway trust fund.

By the end of the 1960s, stories about delays at major airports were a staple of the evening television news and front page articles in major newspapers. Beginning in 1968 the FAA chose to intervene by instituting capacity controls at four airports: Chicago O'Hare (ORD); New York Kennedy (JFK); New York LaGuardia (LGA); and Washington Reagan National (DCA).[33] The high density rule (HDR), as it was known, instituted controls by allocating slots (takeoffs or landings) amongst the incumbent airlines serving each airport. When the slot controls were implemented there was some hope that they could be removed at some not too distant point in time. Instead, the controls would last for over 3 decades. The slot controls system, which was modified on several occasions over the years, probably did constrain excessive growth in congestion at the 4 airports, but it did not eliminate it.[34]

From 1968 until 1986 the allocation of slots was primarily handled by a scheduling committee at each airport made up largely by representatives of the incumbent airlines. The scheduling committees essentially allocated slots amongst themselves on a voluntary basis. This somewhat cozy arrangement was viewed by many as a major restraint on competition. Not surprisingly the scheduling committees failed to provide many slots to the "new entrant" airlines being created in the wake of airline deregulation in 1978. To the extent that new entrants did get slot awards they were provided through intervention by the FAA.

In 1986 an attempt was made to use market forces to open up some of the slots at the airports. This so called "buy/sell" rule allowed incumbent airlines the right to buy or sell slots in a secondary market. To insure that incumbent airlines actually gave up their slots the FAA established minimum use rule for slots. If the slot was not used over a period of time the airline was supposed to lose control over it and return it to the FAA. In reality the system proved far less useful than hoped.[35] Incumbent airlines proved largely unwilling to give up what they regarded as a controlled asset, especially to would-be competitors. Sales of slots, did occur, sometimes as a result of an airline merger, restructuring, etc., but largely in a way that led to even greater control of the HDR airports by major airlines.[36]

By the end of the 1990s it was clear that slot markets had not worked as hoped. FAA intervention over the slot control period, mostly through the award of exemption authority for certain airlines to provide certain types of services, also was seen as having only limited positive impact on competition at the HDR airports. In the end, it was this concern about a lack of competition that led Congress to legislate a phase out of slot controls at all but Washington Reagan National, as part of the Wendell H. Ford Aviation Investment and Reform Act for the 21st Century (AIR-21, P.L. 106-181).

AIR-21 called for a all slot controls to be eliminated at Chicago O'Hare by July 2002, and controls at New York airports were to end January 1, 2007. The law also provided certain exemptions from these deadlines for certain types of aircraft. The first consequences of this phase out occurred at LaGuardia in 2000 when airlines took advantage of small regional jet exemptions provided for in AIR-21 and rapidly initiated a large number of new flights to and from the airport. This jump in regional jet traffic was somewhat unexpected and the FAA was forced to impose a moratorium on new service in September 2000. The events of September 11th temporarily alleviated what was then viewed as a growing problem, but the problem returned and capacity controls were reintroduced by the FAA.

At Chicago O'Hare, the FAA has been able to manage capacity since slot controls were eliminated only by seeking voluntary reductions in scheduled traffic from the incumbent airlines, primarily American and United. In late summer 2006, the FAA adopted a rule that limits operations at the airport until new airport capacity becomes available, which is expected to occur in 2009.

Unlike LaGuardia and O'Hare, JFK did not see the same rapid growth in traffic after passage of AIR-21. For part of this period this was a result of decreases in overall traffic after September 11, 2001. In the last couple of years, however, there has been a major increase in traffic at JFK to the point that the FAA is now seeking to implement some sort of capacity controls at the airport. Options being explored include voluntary rescheduling actions on the part of the airlines or implementation of peak period and/or congestion pricing.

At the time of this writing the FAA seems intent on limiting JFK flights on a per hour basis.[37] The FAA will attempt to meet this goal by asking individual airlines to limit their service at the airport accordingly, but apparently will not allow the airlines to meet to discuss flight scheduling en mass. Longer term, however, the FAA still seems to be focused on creating a congestion pricing scheme to allocate capacity at the airport.[38]

Antitrust Immunity

On a few occasions over the last three decades, the now sunset Civil Aeronautics Board and later the DOT gave airlines limited antitrust immunity in order to discuss and coordinate schedules. Each of these grants was temporary in nature and related to specific events in the aviation industry, such as the aftereffects of the 1981 air traffic controllers strike. In early 2001, after the aviation system experienced massive delays in 2000, Congress was considering legislation (H.R. 1407 and S. 633) that would have allowed airlines to ask the Secretary of Transportation for authority to discuss schedules when scheduled traffic exceeded airport capacity and gave the Secretary the authority to approve the resulting agreements. The authors of these bills viewed the provisions as a short term solution to existing capacity problems, which they hoped would be alleviated over time by the creation of new airport and ATC capacity. The events of September 11, however, would abrogate the need for this legislation at that time.

Early in 2007 concern about congestion at LGA and other airports caused the Bush Administration and the House Committee on Transportation and Infrastructure to include provisions in FAA reauthorization legislation under consideration at that time, H.R. 1356/S. 1076 and H.R. 2881, respectively, to deal with this issue. The Bush Administration approach seeks to study congestion pricing, which will be discussed subsequently. H.R. 2881, however, contains a provision which implies a process that seems similar in some respects to a limited grant of antitrust immunity. In Section 422, the FAA Administrator is given the authority to seek a voluntary meeting of air carriers at a congested airport for the purposes of schedule reduction. If this meeting is unsuccessful, the Administrator is given the authority to facilitate schedule reduction as he/she sees fit.

A similar conversation has now begun in conjunction with the JFK situation. Several parties have called on the FAA to be more active in schedule coordination. For the moment, the FAA has not sought antitrust immunity and is meeting with air carriers one at a time in an attempt to reduce flights into the airport next year.[39] Should this effort fail to achieve its desired results, it is quite possible that the various parties to the scheduling debate might ask for a process that includes at least some form of antitrust immunity. In 2001, there was considerable concern from some observers that antitrust-created scheduling committees or mechanisms would be unable to avoid competitive issues that might lead to industry collusion on items such as service and fares. There also were concerns that scheduling committees, like those used at HDR airports, might become a semi-permanent and anti-competitive feature of the industry.

Congestion Pricing/Peak Period Pricing

In concept the idea of congestion pricing is quite simple, for example, it may consist of modified landing and/or air navigation fees to change behavior in a way that uses the market to allocate available infrastructure. It is the application of the concept in the aviation sector that has proved difficult. To a large extent the difficulty arises because airport runways and air navigation (ATC) services within the United States are normally provided by the public sector and are regarded as public goods. Compounding this problem is the fact that airports are usually local and/or regional monopolies, while ATC services are provided by a national monopoly (such as the FAA in the U.S. context). Pricing in the public sector is, therefore, not focused on creating a profit as it would be for a private firm, but rather on providing for cost recovery or some other publically determined goal. Hence, the forces that routinely set prices in a competitive marketplace are largely absent.

The concept of congestion pricing predates the invention of the airplane. Tolling, which in certain instances is a form of congestion pricing, has been around for centuries. Over the last several decades there has been considerable research on the usefulness and application of congestion pricing mechanisms to transportation infrastructure worldwide. Most of this research, however, has focused on surface transportation. In the surface transportation world, there is a large and growing body of academic literature on the subject. The similar conversation in the aviation world is much more limited in scope, with research on the subject being done by a comparatively small number of individuals and groups. And, in fact, several of the authors of studies on aviation pricing are doing so based, at least in part, on their research on surface transportation pricing issues.

The Bush Administration, and especially Secretary of Transportation Peters, favor market mechanisms for infrastructure development and maintenance in all transportation modes.[40] It was, therefore, not a surprise that the Administration made a review of congestion pricing part of its proposal for FAA reauthorization in early 2007, H.R. 1356/S. 1076. The FAA proposal seeks to establish a pilot program to evaluate market-based mechanisms to relieve congestion at up to 16 airports (including LGA, which was the subject of a separate provision). The FAA proposal was not adopted in either the House (H.R. 2881) or Senate (S. 1300) reauthorization bills still under consideration at the end of 2007.

Aviation Infrastructure Pricing

Airports in the United States currently charge landing fees based on aircraft weight. These are generally fixed and do not vary over the time of day. With a couple of small exceptions, there is no correlation between the landing fee and congestion, although there may be some correlation between fee levels and infrastructure needs at the airport. Typically, in the absence of specific infrastructure needs, airport landing fees are determined on the basis of the historical/administrative costs associated with the airport's provision of service to the airlines and its own operating cost requirements. While landing fees are an important source of revenue for most airports, they are not the only source. And at many airports landing fees are not even the majority source of airport income. At very large airports, for example, the airport may receive far more income from concession fees than it does from landing fees and other airside related activities.

U.S. and Non-U.S. Experience With Congestion Pricing

Congestion pricing is an idea that has not yet come to fruition in the aviation world for a variety of reasons, some of which will be discussed below. As a result, the body of experience with congestion pricing is quite limited, having been applied in some form only at New York's and Boston's major air carrier airports, and at London Heathrow (LHR) airport in the United Kingdom (UK). In no case can the application of these fees be viewed as having been entirely successful from the perspective of congestion pricing theory. And in no case can any of these examples be viewed as having created a system that would meet the theoretical definition of true congestion pricing. Each of these experiences is discussed briefly below.[41]

New York

In 1968, the Port Authority of New York and New Jersey (Port Authority) instituted a $25 fee/surcharge on all aircraft operations for aircraft with 25 seats or less during peak hours at the three air carrier airports it operated.[42] The specific intent of this first known instance of peak period pricing in the United States was to reduce GA operations at the three airports. The hope was that GA operations, especially those using LGA, would move to Teterboro Airport (TEB), which was also operated by the Port Authority and had no scheduled air service. No additional fees were to be imposed on GA aircraft that chose to use TEB as a substitute. Not coincidentally, this action took place during the same time period that saw the implementation of the HDR slot controls discussed earlier.

GA interests attempted to overturn these surcharges in the federal courts but were ultimately unsuccessful in doing so.[43] The Court found that the intent of the surcharge was reasonable and that the Port Authority had not discriminated against GA. In part, this was because the Court held that TEB provided a reasonable alternative for GA aircraft that wished to avoid the surcharge. The peak period surcharge system at LGA remains in place today.

The peak period pricing system using landing fee surcharges was successful in moving the majority of GA traffic out of LGA and into TEB. In this respect, the Port Authority experience can be viewed as having been successful. It should, however, be recognized that the goals of the surcharge were modest when compared to the peak period pricing regimes promoted by economic theorists. Hence, it would be difficult to draw on this experience as a guide to how peak period surcharges might affect the movement of airline traffic in the context of the ongoing FAA/DOT slot allocation investigation at JFK.

Boston

Some would argue that Boston Logan's 1988 attempt at using pricing to modify airport traffic flow was not, in fact, an example of congestion pricing, *per se*. The Program for Airfield Capacity Efficiency (PACE) changed the landing fee structure at the airport by moving away from a strictly weight-based fee and instead creating a base fee unrelated to weight for all aircraft landings with an additional weight-based fee. The net effect of this system was to significantly raise the landing fees paid by GA aircraft and smaller commuter/regional airline aircraft as compared to those now paid by larger airline operated aircraft. Massport, the state agency operating the airport, believed that this pricing system would reduce congestion by providing an incentive for airlines to schedule fewer flights during the day using larger aircraft and cause GA and other aircraft to fly to other Boston area airports. In short, Massport was attempting to use landing fees as an upguaging strategy.

GA and the commuter/regional airlines faced with higher fees mounted a legal challenge to the PACE program. The complicated legal battle that ensued is discussed in more detail later in this chapter. For the purposes of this discussion only the result of that process is addressed. The final decision in the case was made by an Administrative Law Judge (ALJ) at DOT who decided that the PACE program discriminated against GA and commuter/regional airlines for several reasons.[44] Important among these was the lack of an acceptable alternative airport for the users being charged the higher fees. Interestingly, however, the ALJ specifically protected the right of the airport to adopt a landing fee system different from the weight based system in place at a later date.

> There is nothing inherently sacrosanct about weight-based fees, and nothing in the law that precludes Massport from changing its method of cost allocation.[45]

As a result of the court action, the PACE program was only in place for six months and it is not clear that the reduction in flights during that period can be specifically attributed to the PACE program.[46] In the intervening years Massport has adopted, but not implemented, a peak period surcharge methodology that it believes complies with DOT's requirements. The surcharge itself will only be implemented when traffic delays reach predetermined levels.

London

London Heathrow (LHR) is one of the world's most congested airports. In an attempt to modify traffic flows it has imposed a peak period pricing regime since 1972. At that time the airport created a new landing fee structure that included flight length, passenger load, time-of-day and weight-based elements. The fees also were related to the individual airline's ability to pay the fee. Over time the specific elements of the LHR fee structure have evolved. For example, after numerous protests by international carriers the distance related portion of the fee was eliminated in 1978. To the chagrin of all air carriers at LHR the British Airports Authority (BAA) which operated the airport responded by dramatically raising the fixed element of the fee so that many air carriers actually ended up paying higher fees than they had previously.

BAA was a public entity in the 1970s when the peak period fees were implemented. Since 1987 it has been a private firm. In the late 1980s, U.S. air carriers upset over the complicated and high landing fees at LHR, successfully sought international arbitration over the fees. The findings of that process concurred with the U.S. air carrier view and BAA was required to reach a financial settlement with the carriers.[47] Irrespective of the U.S. air carrier experience, other elements of the landing fee structure have continued to change, but the single constant over time has been a consistent increase in the landing fees.

LHR is a heavily congested slot controlled airport with much of the capacity (slots) at the airport reserved as a result of international aviation agreements. Therefore, there are other mechanisms in place at the airport besides peak period pricing that allocate capacity to some degree. Airline access to LHR is one of the most sought after commodities in the world of aviation. It is, therefore, an airport that nearly all international airlines regard as a desirable destination, regardless of the high landing fees and other costs associated with its use. Not surprisingly, major airlines have continued to seek additional service opportunities at the airport. This has not been the case for commuter/regional airline service that has largely moved to alternative London airports such as Stansted.[48]

Most observers would suggest that the London example is the closest to true congestion pricing and has had some of the types of results that one might theoretically expect from its application. The reality, however, is that LHR's peak period fees have never really created a market-based allocation of traffic amongst major airlines. This is due primarily to the desirability of LHR as a destination. Regardless of the high fees at the airport, airlines appear to have been able to transfer all of the costs associated with the peak period pricing regime to their passengers through higher fares. Interestingly, congestion is such at LHR that there is no longer a true peak period and the landing fees have been modified to recognize this reality.

Issues

There are a number of issues that will need to be considered as part of any consideration of congestion pricing. Each of these issues is potentially contentious. What follows is a brief discussion of several of these issues.

Fee Setting

Although there are other airport related charges that might be modified or created to effect congestion pricing at an airport, experience to date clearly shows a preference for using landing fees as the principal mechanism for price setting. In the United States, the setting of landing fees has been the domain of the airport operator. The legal discussion that follows this section will show that landing fees can be challenged on various grounds, but nonetheless airports have wide discretion in setting these fees so long as they do not discriminate against specific users and the methodology employed in setting the fees is considered fair. As a result, airport operators might contend and have, in the past, argued that they should determine the pricing policy to be used. Airport proponents might also argue that they can also effect the rationing of airport capacity by imposing fees on other types of airport airside infrastructure such as terminal gates.

Although most of the discussion to date about congestion pricing has been focused on the use of a specific airport, or groups of airports, it is possible to consider congestion pricing of the airspace in the proximity of the airport or region as an alternate pricing mechanism. In this instance a user fee might be set by the FAA for access to an airport's related airspace. TRACON (terminal radar approach control)- related user fees were part of the FAA's 2007 reauthorization proposal (H.R. 1356/S. 1076). These proposed fees were not seen as congestion pricing fees, however, but rather as a cost recovery mechanism.

Fee Setting and Use Issues

Two important issues that will be raised in any congestion pricing environment are the related questions of what kinds of outcomes the fees are to achieve and how any money collected is to be used. If the goal is true congestion relief the fees might have to be set quite high in order to prompt airlines to make schedule changes. One possible result of high fees, however, is that certain user groups such as regional airlines providing Essential Air Service (EAS) or other services to smaller communities might be priced out of specific markets. This result, and other similar types of exclusions, could be at odds with the existing public policy goal of providing broad access to the national aviation system.

How collected fees might be used also has potential implications that could determine whether or not congestion fees would enjoy public and/or airline industry acceptance. For example, the use of landing fees to fund the operating costs of an airport and to pay for the creation of airport infrastructure is broadly accepted as an appropriate use for these funds. This might not be the case if fees were instead designated for U.S. Treasury General Funds without any specific designation for aviation purposes.[49]

Also in question is how congestion fees might affect airline fares at an individual airport. Airlines ultimately transfer their costs to their passengers. It is well understood that airline fares to and from airports where a single air carrier dominates the market are higher than they are at airports with high levels of competition. Congestion fees could result in greater concentration at certain airports giving air carriers with the majority of service at an airport significant new market pricing power. Unknown is whether congestion fees could result in high airline fares leading to public unhappiness with the resultant capacity rationing system. This especially would be the case in markets where substitute airports and/or other transportation alternatives were lacking. On the other hand, if high fees did result in significant reductions in air carrier delay and improved flight reliability, the public might accept the higher fares.

Alternative Fee Mechanisms

Since the HDR process was first implemented there have been numerous attempts to create market forces for their allocation. As discussed earlier, most of these attempts were seen as not having fully served their intended purposes. Some researchers, however, have suggested that these airport-centric mechanisms have not embraced the market sufficiently and that slot-related charges are still a viable option for the allocation of airport and airspace capacity.

An approach suggested by some researchers is the slot auction. In one instance the researchers suggest a multi-level auction process to allocate capacity at congested airports using long-term leases of slots at congested airports (one year or so), with a secondary market that allows airlines to trade these leases, and an additional market that allows airlines to trade these resources amongst themselves on a day-to-day basis.[50] In this particular instance the market would likely be run by the FAA with funds derived from the slot auctions reserved for airspace infrastructure improvements. The researchers recognize a significant number of issues that might arise from this process, but believe it could be a useful mechanism for the allocation of capacity.[51]

Legal Issues

The consideration and potential imposition of congestion pricing or other related mechanisms at airports raises numerous legal issues and highlights a long-standing tension in aviation law between federal uniformity and supremacy over airspace and air safety concerns on the one hand, and local control over airports and ground facilities on the other. This tension, in addition to legal issues related to the ability of airports to use congestion pricing or similar mechanisms in conjunction with the federal financial support received through programs such as the Airport Improvement Program (AIP), may have a significant impact on the type of mechanism suggested and considered.

As noted earlier, it appears that the legal issues related to congestion pricing mechanisms have been primarily identified by the two local airport authorities, who sought to impose such systems over the legal objections, generally on preemption grounds, of the FAA and other adversely affected parties. To date, the FAA has not suggested the imposition of a federal congestion pricing mechanism on local airports. As a result, it is unclear precisely what, if any, legal issues may arise should the situation reverse itself and congestion pricing become a federal mandate as opposed to a local solution seeking federal approval.

This section of the chapter will proceed as follows. First, it will review and survey the current laws that govern airports, including the federal preemption statute as well as several other federal laws that constrain what local airports are permitted to do with fees charged and money granted by federal programs such as the AIP. Second, the chapter will discuss current law as it relates to airport's proprietary powers. In other words, given the federal statutory structure, what authority, if any, is left to the local governments and airport authorities with respect to implementing congestion pricing mechanisms? Third, the chapter will review the two previous examples of alternative pricing mechanisms proposed by local airport authorities, both of which resulted in federal litigation. The first, in 1969, at New York's LaGuardia Airport, was approved by the federal court, while the second, at Boston's Logan Airport, was rejected by the reviewing court. Finally, this section will discuss some of the legal issues that may arise should the FAA attempt to offer congestion pricing as a federal mandate. These may include such issues as whether the FAA has the proper statutory authority and whether laws related to the financing and use of funds by airports may need to be adjusted or amended.

Federal Aviation Laws

Since the passage of the Civil Aeronautics Act in 1938 (CAA), [52] the federal government has worked to establish a coherent national aviation policy that places the primary responsibility for regulating the air transportation system with federal agencies, while at the same time attempting to retain local control over certain ground-related elements, including monitoring of noise, land-use, and zoning. The CAA set forth, *inter alia*, the encouragement and development of the air transportation system, the regulation of air transportation to assure a high degree of safety and sound economic conditions, and the promotion of efficient service by air carriers without unfair competitive practices or unjust discrimination.[53] Although replaced in 1958 by the Federal Aviation Act,[54] the general policy language has largely been retained and the federal government has continued to assert primary control over much of the aviation sector, especially with respect to airspace, safety, and discrimination related issues. In addition, the Federal Aviation Act included a specific preemption provision that prohibits state and local authorities, including airport authorities, from enacting and enforcing laws related to airline "price, route, or service."[55] Moreover, the statutory language expressly preserves state and local authority with respect to an airport's "proprietary powers," which will be discussed in more detail below.[56] Finally, it should be noted that most economic regulation of the aviation sector was eliminated by the adoption of the Airline Deregulation Act of 1978.[57]

With respect to airspace, current law clearly establishes that it is the federal government that controls the sovereignty of U.S. airspace.[58] Specifically, the Federal Aviation Administration (FAA) is charged with "develop[ing] plans and policy for the use of the navigable airspace and assign[ing] by regulation or order the use of the airspace necessary to ensure the safety of aircraft and the efficient use of airspace."[59] Statutes such as these are generally considered "field preemption" statutes or instances where Congress, by the nature of its action, can claim to have preempted or occupied an entire area of the law. Generally, the Supreme Court has found field preemption in two types of situations: (1) where "the scheme of federal regulation is sufficiently comprehensive to make reasonable the inference that Congress 'left no room' for supplementary state regulation;"[60] or (2) "where the field is one in which 'the federal interest is so dominant that the federal system will be assumed to preclude enforcement of state laws on the same subject.'"[61] As a result of these statutes, the states and other political subdivisions such as local governments and airport authorities have had limited ability to adopt measures and policies that interfere with the federal government's ability to control airspace and other aviation related issues.[62]

In addition to the federal preemption statutes, local governments and airport authorities are also limited by federal law with respect to their ability to raise and spend specific fees and taxes. For example, taxes and other charges imposed by local governments or airport authorities for use of the facility are generally prohibited by the Anti-Head Tax Act of 1973.[63] Moreover, the Anti-Head Tax Act prohibits state and/or local governments, and airport authorities from taking other actions that "unreasonably burden and discriminate against interstate commerce."[64]

The acceptance of federal funding, specifically those awarded out of the Airport Improvement Fund (AIP), also restricts local authority with respect to aviation issues. Specifically, as a condition of receiving AIP grant funds, an airport must provide access to its facilities on a reasonable basis and without discrimination, and must also charge airlines similar prices for similar use.[65] Furthermore, the statute requires that airports accepting AIP funds must undertake efforts to ensure that its rate structure makes the facility "as self-sustaining as possible."[66]

Airport's Proprietary Powers

In *City of Burbank v. Lockheed Air Terminal, Inc.*,[67] the Supreme Court, in an opinion by Justice Douglas, established what has become known as the "proprietor exception" to the general federal preemption of local authority with respect to aviation law. In creating this exception, which reserves some authority for local governments and airport authorities, Justice Douglas appeared to draw a distinction between municipalities exercising control as proprietors and municipalities exercising police powers in a way that conflicts with federal law.[68]

Since *City of Burbank*, the lower federal courts have generally held that "an airport proprietor can issue only 'reasonable, nonarbitrary, and nondiscriminatory rules that advance the local interest.'"[69] While under this standard courts have upheld regulations aimed at monitoring noise levels,[70] addressing environmental concerns,[71] and managing ground congestion,[72] there appears to be no justification for a broad grant of proprietary power. In 2000, the Fifth Circuit rejected an interpretation of the proprietor exception that would have permitted an airport owner to "allocate traffic between two airports so as to preserve the short-haul nature of one facility."[73] It is worth noting, however, that in the same Fifth Circuit Court of Appeals decision, which involved the now-repealed "Wright Amendment," the court did note a willingness to review the exercise of proprietary powers with respect to advancing previously unrecognized local interests, but not in this particular case.[74] As none of these rulings appear to have addressed the issue of congestion pricing, it is at best unclear whether such authority is permissible or within the so-called "proprietor exemption."

Previous Attempts to Impose "Congestion Pricing" Mechanisms

As previously mentioned, there have been two attempts at imposing fee mechanisms that might be considered "congestion pricing" broadly defined. These attempts both emanated from the local airport authorities and were met with resistence from adversely affected interests and, in the case of Boston's Logan International, from the FAA, primarily on federal preemption grounds. We will discuss each attempt in turn.

New York's LaGuardia Airport

In 1968, the Port Authority of New York and New Jersey, which operates four airports in and around New York City (LaGuardia (LGA), John F. Kennedy International (JFK), Newark Liberty International (EWR), and Teterboro (TEB)), sought — "for the purpose of relieving congestion and achieving the maximum efficient operation at the airports" — to impose a $25 minimum charge for landing during peak hours.[75] The fee applied to any aircraft that took off or landed during the peak hours and had a seating configuration of less than 25 passengers.[76] The fee applies at the three major Port Authority air-carrier airports, LaGuardia, JFK, and Newark, but not at Teterboro Airport. A lawsuit was brought on the grounds that the so-called "take-off"fees impermissibly discriminated in favor of commercial airlines and against general aviation interests.

Plaintiffs argued, *inter alia*, that the FAA is vested with the sole authority to regulate and control the national airspace and, therefore, the Port Authority violated federal law in imposing the $25 "take-off" fee.[77] In addition to the preemption argument, plaintiffs argued first that the FAA is required by law to expressly approve of the Port Authority's regulations, or at a minimum is required to conform their compatibility with grant and other aid agreements.[78] Second, according to the plaintiffs, "Teterboro is not a reasonably convenient nor adequate facility as an alternate to the three major New York airports" and, therefore, the diversion of general aviation to Teterboro would result in a practical exclusion from the New York area airports. Finally, the plaintiffs asserted a variety of non-legal or policy arguments against the proposal. These arguments included that it would be ineffective to remedy the airport situation, which was due to over-scheduling by commercial airlines and not general aviation,[79] that diversion would not resolve the disparity between airport capacity and demand,[80] and that the record indicated that the commercial airlines were intimately involved in the proposal and even offered to subsidize ground transportation from Teterboro to provide further incentive for moving general aviation away from the other airports.[81]

In response to the preemption argument, the court first noted that there exists room "for the operation of Port Authority Regulations which have the effect of curtailing activities not forbidden by federal regulation."[82] As a result, according to the court, nothing in the proposed Port Authority regulations "runs counter to the FAA regulation in the sense that it seeks to authorize conduct which the federal regulation prohibits or requires the cessation of a practice required by federal regulation."[83] In response to the argument that the FAA was required to approve the Port Authority regulations, the court held that there was enough evidence on the record to indicate that there was no "inherent incompatibility" with federal law.[84] Next, the court addressed the argument that Teterboro is not "reasonably convenient" to New York City. The court held that Teterboro is "not a nil facility or a purgatorial alternative" and, thus, does provide a valid basis for limiting the Port Authority's attempts to relieve the region's congestion.[85] With respect to the other arguments raised by the plaintiffs, the court found no facts relevant to the dispute and entered judgment in favor of the defendants, thereby allowing the fees to go into effect.[86]

Boston's Logan International Airport

In 1989, the First Circuit Court of Appeals decided *New England Legal Foundation v. Massachusetts Port Authority*,[87] which addressed the validity of the landing fee scheme implemented by the Massachusetts Port Authority (Massport) as part of a multi-phased "Program for Airport Capacity Efficiency (PACE)."[88] The fees at issue were to be a combination of a flat-rate landing fee of $91.78 plus a weight-based fee of $0.5417 per 1000 pounds.[89]

Several groups that opposed the new landing fee structure filed an administrative complaint with the FAA alleging that the fees were invalid because they were not "fair and reasonable" and they discriminated against smaller aircraft.[90] The Secretary of the Department of Transportation (DOT) agreed to review the complaint and commenced an investigation and requested that Massport suspend implementation of PACE.[91] Massport declined and indicated that it would implement the new fees on July 1, 1988. While the FAA investigation was pending, a separate group filed suit in federal court seeking declaratory and injunctive relief preventing Massport from implementing the PACE landing fees.[92] In addition, the opposition groups who had sought administrative relief also filed suits in federal court. All three lawsuits contained similar allegations and were consolidated. The suits alleged that by shifting the cost burden of the landing fees from the larger aircraft to the smaller, the new method conflicts with various federal statutes that regulate the national airspace and were therefore preempted.[93] The fees were also alleged to be constitutionally invalid not only because by placing an undue burden on interstate commerce they violated the Commerce Clause,[94] but also because they discriminated illegally against small aircraft in contravention of the Equal Protection and Due Process Clauses of the Fourteenth Amendment.[95]

The district court, ruling orally from the bench, found in favor of Massport. On the question of preemption, the court found that there was no preemption because Congress, in passing the Airline Deregulation Act of 1978,[96] intended the proprietor to set the landing fees.[97] With respect to allegations of unreasonableness, the court found that the fees were reasonable because they were fixed "according to [a] uniform, fair and practical standard ... [and were] non-excessive in comparison with the governmental benefit conferred" and were "relevant to the operation of the airport ... and fairly consonant with the cost incurred."[98] Finally, regarding the alleged discriminatory effect, the court held that "because the objectives of the airport to reduce congestion and delay are rationally related to legitimate governmental interest ...[the fees] enhance economic and operating efficiency" and, therefore, are not discriminatory.[99] As a result of the district court's ruling, the PACE fee structure went into effect on July 1, 1988.

Litigation then shifted to the DOT, who appointed an Administrative Law Judge (ALJ) to recommend a decision based on the evidence submitted. The ALJ concluded that the PACE fees were: (1) in violation of Massport's grant assurances as they were not "fair and reasonable;" (2) unjustly discriminatory, and were in violation of various federal statutes,[100] and thereby invasive of the DOT's authority and; (3) preempted by federal law.[101] In addition, to these major findings, the ALJ also held that Hanscom Field cannot be considered an alternative to Logan International because it "lacks the connecting opportunities that are otherwise available at Logan."[102] On appeal to the Secretary of Transportation, the Secretary affirmed and adopted the ALJ's recommendations; however, the Secretary indicated that the ALJ "overstated the extent of the preemptive role of the federal government in the control of airport access and access to the national air transportation system."[103] According to the Secretary, "it is ... within an airport proprietor's authority to impose reasonable [nondiscriminatory,] landing fees or other user fees, even if such fees may result in a decline in usage by a class of user or other indirect effects on users."[104] Moreover, "[w]hile an airport sponsor has wide latitude in recovering costs and using fees to improve the efficient use of airport facilities, the FAA is vested with the authority to control, regulate and manage air traffic."[105] As a result of the Secretary's decision, Massport suspended the PACE fee on December 27, 1988, pending the outcome of its appeal.

Both the district court's decision and the Secretary's affirmation of the ALJ's findings were appealed to the First Circuit Court of Appeals. The appeals court divided its analysis by statutory provision. First, the court addressed the allegations that the fees were unreasonable and in violation of section 511 of the Airport and Airway Improvement Act of 1982. The court affirmed the Secretary's determination that the fees were unreasonable on agency deference grounds, concluding that the record firmly supports the Secretary's conclusions and that the issue is "definitely an area of expertise primarily reserved to the Secretary's discretion, and not to be second-guessed by appellate courts...."[106]

Next, the court reviewed the findings of both the district court and the Secretary with respect to the Federal Anti-Head Tax Act. The district court had found the fees to be reasonable, thus not in violation, while the ALJ and ultimately the Secretary had concluded that the fee was not a head-tax or its equivalent, thus the statute was not implicated. The appeals court upheld both holdings, though it utilized the rationale provided by the Secretary, concluding that Massport's fee structure was "outside the scope of [the Act], as not being ... 'a charge ... on persons traveling in air commerce,' nor is it a levy 'on the carriage of persons' so traveling, 'on the sale of air transportation,' or 'on the gross receipts derived therefrom.'"[107]

Finally, the appeals court addressed the preemption issues raised by sections 105 and 307 of the Federal Aviation Act of 1958 as amended. On this issue the district court and the Secretary had disagreed, with the court finding the fees reasonable and hence not preempted, while the Secretary found them to be unreasonable and, therefore, preempted by federal law. The appeals court ultimately concluded that the district court erred "not only in its decision, but also in not deferring to the Secretary's primary jurisdiction over this controversy as was requested by the DOT in its last minute amicus intercession." Relying on the doctrine of "primary jurisdiction" the court held that in cases such as these "[w]hen there is a basis for judicial action, independent of agency proceedings, courts may route the threshold decisions as to certain issues to the agency charged with primary responsibility for governmental supervision or control of the particular industry or activity involved."[108] In other words, the district court should have deferred its judgment until after the FAA and the Secretary had investigated and completed their administrative review. Turning to the merits of the Secretary's decision, the court concluded that Massport's fee structure was an impermissible interference with air navigation and, therefore, preempted by federal law.[109] Having decided the case on statutory grounds, the court declined to address the constitutional issues raised by the Commerce Clause or the Fourteenth Amendment.[110]

Based on these two cases it is difficult to discern precisely where the courts are with respect to airport imposed pricing mechanisms intended to relieve congestion. At one level it seems that perhaps the different results reached turned on the availability and feasibility of alternative landing sites for the smaller general aviation planes that were adversely affected by both pricing schemes. In the case of New York and Teterboro, the court found that to be a viable alternative, whereas with respect to Boston and Hanscom the ALJ was not persuaded that the alternative location was reasonable. Resting on this distinction, however, appears to oversimplify many of the complex issues that the court and the ALJ dealt with in the Massport case. The development of aviation law and addition of new statutes in the 20 years between the two cases appeared to play an integral role in the difference between the results.

Currently, the law remains nearly identical to what it was in *Massport*, thereby making it unlikely that another attempt by a local airport authority to impose a similar pricing scheme without the FAA's express approval would succeed. Admittedly, this analysis appears to render the fees at LaGuardia *sui generis*;[111] however, such a conclusion appears to be a reasonable way to reconcile what on their face seem to be two irreconcilable judicial decisions. That said, it does not appear to be *per se* impossible for local airport authorities to take action. It may be that a fee structure could be developed that imposes fees — designed to reduce congestion — that are reasonable, non-discriminatory, and based on the airport's historical operating costs. Assuming such a fee structure could be developed, it appears possible that it may not be in violation of either federal law or FAA regulations.

Potential Legal Issues Should FAA Impose Congestion Pricing

As the previous section indicated, all of the prior attempts at imposing congestion pricing mechanisms have been locally generated by the airport authorities. Neither the FAA nor the DOT have, to date, attempted to develop and impose a congestion pricing scheme on the local authorities. In the event that FAA should decide to make such an attempt, it appears that there are several legal issues that may arise.

First, there appears to be a question of legal authority. Does FAA have the legal authority to impose such a system without a change in the current statutes? Based on a review of the statutes currently in place, it appears that the FAA has a reasonable basis to proceed. While it is true that there is no specific statute that directly provides FAA with such authority, an argument can be made that, by virtue of the agency's statutory mandate with respect to the airspace of the United States, the legal authority exists. Phrased another way, if the FAA characterizes the imposition of a congestion pricing scheme as an "airspace," "air traffic," or "air navigation" means of managing it may be argued that it would fall within the agency's existing authority provided by Congress to "develop plans and policy for the use of the navigable airspace and assign by regulation or order the use of the airspace necessary to ensure the safety of aircraft and the efficient use of airspace."[112] This argument would also appear to be consistent with the Secretary's finding in *Massport* that the "FAA is vested with the authority to control, regulate and manage air traffic."[113] Yet another way of looking at this is potentially through a "negative preemption" type argument.[114] In other words, if, as the DOT argued in *Massport*, the local airport authorities are preempted by federal law from imposing unreasonable and/or discriminatory fees — even to reduce congestion — then it stands to reason that the federal government, through the FAA, is the *only* place where such a regulation could originate.

In addition, it should be noted that there appears to be a generally-applicable federal statute that permits the heads of agencies to "prescribe regulations establishing the charge for a service or thing of value provided by the agency."[115] The statute requires that each charge "shall be: (1) fair; and (2) based on — (A) the costs to the Government; (B) the value of the service or thing to the recipient; (C) public policy or interest served; and (D) other relevant facts."[116] The Supreme Court has held that this authority extends to "only specific charges for specific services to specific individuals or companies."[117] Based on the plain language of this statute, it appears possible for the FAA to assert that the "thing of value" provided is access to the national airspace; therefore, imposing fees and charges, provided they meet the requirements of the statute, is arguably permissible. Although it does not appear that this statute has ever been used with respect to fees relating to airspace use, during its existence the Civil Aeronautics Board did use the statute to impose and modify various filing and licensing fees,[118] so its use in the aviation context is not without precedent.

A second question that arises has to do with the idea of reasonableness and non-discrimination in setting airport fees. If one accepts the authority argument that only the FAA can impose a congestion pricing system, it appears to follow then that the restrictions currently in place that prohibit state and local authorities from discriminating and imposing unreasonable fees would not apply, as the pricing scheme would be federally imposed and not locally mandated. Similarly, none of the AIP restrictions on local airport authorities appear to be implicated by an FAA imposed congestion pricing scheme. A more complicated set of questions arise should the FAA decide not to impose congestion pricing directly, but rather to delegate the authority to impose congestion pricing to various local airport authorities. In this situation, it would appear that either FAA would have to waive the restrictions on reasonableness and non-discrimination or simply refuse to enforce the potential violations that may ensue from implementation of its mandate. Based on a plain reading of both the Anti-Head Tax[119] and the Airport and Airway Improvement Act of 1982,[120] neither statute appears to provide the Secretary or the Administrator of the FAA with the authority to grant waivers for states to levy taxes, or from the non-discrimination and reasonableness requirements. In fact, the Airport and Airway Improvement Act of 1982 seems to require that approval be granted "***only if*** the Secretary receives written assurances, satisfactory to the Secretary, that ... the airport will be available for public use on reasonable conditions and without unjust discrimination."[121] Hence, it would appear that should the FAA attempt to delegate its authority over a potential congestion pricing scheme to the local airport authorities, it will first have to seek a statutory change from Congress.

A third question arises with respect to the use of any funds collected from the imposition of a federally mandated congestion pricing scheme. Even if it is accepted that FAA has the legal authority to implement a congestion pricing scheme it does not automatically follow that the FAA or the DOT would be permitted to retain the fees collected. It is a generally accepted principle of federal appropriations law that absent statutory authority to the contrary, the "miscellaneous receipts statute"[122] requires agencies to deposit funds received for the government into the federal Treasury, otherwise known as the "general fund." An agency may only retain such moneys it receives if it has statutory authority to do so. According to decisions of the Comptroller General, the "miscellaneous receipts statute" does not apply if there is specific statutory authority for the agency to retain the funds.[123] Consequently, absent such specific authority, it would appear that even if the FAA were to delegate congestion pricing authority to the local airport authorities, the money would still be required to go to the Treasury's "general fund." This conclusion follows from principles of delegation; namely, that government entities cannot delegate authority that they themselves do not possess.[124] Hence, if FAA does not possess the authority to keep the fees that it collects from a congestion pricing scheme, it cannot properly delegate the power to keep the fees to a local airport authority. In sum, should it be desired that the funds collected from a congestion pricing mechanism go anywhere but the "general fund," Congress will have to pass a statute authorizing the retention and expenditures of the funds. Similarly, it appears that any expenditures of funds collected will have to be specifically authorized and appropriated by Congress.[125]

END NOTES

[1] Wall, Matthew L. & Belson, Ken. U.S. to Limit Flights at Kennedy. *The New York Times.* December 11, 2007. p. A31.

[2] See Scott McCartney, "Why Flights Are Getting Longer," *The Wall Street Journal*, May 29, 2007, p. D1.

[3] Ibid.

[4] "White House Unveils Plan to Ease Holiday Travel Congestion," Associated Press, November 15, 2007.

[5] Remarks for the Honorable Mary Peters, Secretary of Transportation, Aviation Congestion Relief Plan, Media Briefing, Washington, DC, September 27, 2007.

[6] The times represent the average (mean) delay for arrivals and departures among flights that were delayed 15 minutes or more.

[7] Statement of Robert A. Sturgell, Acting Administrator, Testimony Before the House Committee on Transportation and Infrastructure, Subcommittee on Aviation, On Airline Delays and Consumer Issues, September 26, 2007.

[8] See Federal Aviation Administration, FAA Aerospace *Forecasts (Fiscal Years 2007—2020)*, Washington, DC: Federal Aviation Administration, Aviation Policy and Plans.

[9] David Hughes, "N.Y.'s Delay Donnybrook," *Aviation Week & Space Technology*, July 30, 2007, pp. 40-42.

[10] George L. Donahue, *Optimum Fleet Utilization Under Congestion Management at NY LGA*, presented at the Workshop on Allocation and Exchange of Airport Access Rights, The Aspen Institute, Wye River Conference Center, Queenstown, MD, June 6-8, 2007.

[11] Ibid. In game theory, the "prisoner's dilemma" refers to a competitive game scenario in which players can collectively gain more by cooperating, but typically compete instead to serve their own self-interests based on their perception that the other player(s) will also compete rather than cooperate.

[12] Federal Aviation Administration and the MITRE Corporation, *Capacity Needs in the National Airspace System, 2007-2026*, May, 2007, p. 14.

[13] Ibid., p. 18.

[14] Ibid.

[15] Joint Planning and Development Office, *Concept of Operations fo the Next Generation Air Transportation System*, Version 2.0, June 13, 2007.

[16] Testimony of Ruth E. Marlin, Executive Vice President, National Air Traffic Controllers Association Before the U.S. House of Representatives Committee on Transportation & Infrastructure, Subcommittee on Aviation, on the Status of the Air Traffic Controller Workforce, June 15, 2004.

[17] Remarks for Robert A. Sturgell, Acting Administrator, Federal Aviation Administration, Aviation Congestion Relief Media Briefing, Washington, DC, September 27, 2007.

[18] Air Transport Association, Letter to the Honorable Mary Peters, Secretary, U.S. Department of Transportation, Washington, DC, September 14, 2007.

[19] See NEXTOR: The National Center of Excellence for Aviation Operations, *Allocation and Exchange of Airport Access Rights*, Workshop held at the Aspen Institute, Wye River Conference Center, Queenstown, MD, June 6-8, 2007, available at [http://www.isr.umd.edu/airworkshop/].

[20] Federal Aviation Administration, *FY2005 Cost Allocation Report*, January 2007.

[21] Federal Aviation Administration, *Terminal Area Forecast (TAF)*, available at [http://aspm.faa.gov/main/taf.asp].

[22] See "Corporate Jets Clog America's Skies," at [http://www.smartskies.org/Learning Center/atc_users/].

[23] Ibid.

[24] Statement of Robert A. Sturgell, Acting Administrator, Testimony Before the House Committee on Transportation and Infrastructure, Subcommittee on Aviation, On Airline Delays and Consumer Issues, September 26, 2007.

[25] Madhu Unnikrishnan, "DOT: Military Airspace Opened For Thanksgiving." *Aviation Daily*, November 16, 2007.

[26] Holly Rosenkrantz and John Hughes, "Bush to Open More Airspace to Cut Holiday Trip Delays," *Bloomberg*, November 15, 2007.

[27] Ibid.

[28] James Fallows, "Thankfulness is Great, But What Is the NYT Thinking?," *The Atlantic.com*, November 22, 2007.

[29] Air Transport Association, ATA 2007 *Thanksgiving Holiday Operational Earlier Updates*, November 17-27, 2007, Washington, DC: Air Transport Association.

[30] Statement of Robert A. Sturgell, Acting Administrator, Testimony Before the House Committee on Transportation and Infrastructure, Subcommittee on Aviation, On Airline Delays and Consumer Issues, September 26, 2007.

[31] "White House Confirms Airline Scheduling Meetings for JFK." *Aviation Daily.* September 28, 2007. p. 1.

[32] Schofield, Adrian. "Debate Over JFK Delays Sparks Talk of Congestion Pricing." *Aviation Week.* October 29, 2007.

[33] Newark Airport was briefly subject to FAA slot controls as well.

[34] For a more detailed description of the history and origin of slot controls, see Gleimer, Eileen M. "Slot Regulation At High Density Airports: How Did We Get Here And Where Are We Going?" *Journal of Air Law and Commerce.* V. 61. May -June 1996. p. 877-931.

[35] Berardino, Frank. "History of the Slot Exchange Market in the US and Some Implications." Presentation at: Nextor Workshop on Allocation and Exchange of Airport Access Rights. June 6, 2007. [http://nextor.org/Conferences/200706_Airport_Access_Rights/Berardino.pdf].

[36] Ibid. p. 10.

[37] "FAA Looks to Cap JFK Flights at 80-81 Per Hour." *Aviation Daily.* October 22, 2007. p. 2.

[38] Ibid.

[39] "FAA Begins JFK Scheduling Talks, But Airlines Slam Flight Cuts." *Aviation Daily.* October 24, 2007. p. 1.

[40] "Transportation Congestion Pricing Would Ease Problems at Airports, on Highways, Peters Tells Panel." *Daily Report for Executives.* BNA, Inc. No. 202. October 19, 2007. p. A-9.

[41] For more detailed information on each example see Schank, Joshua L. "Solving airside airport congestion: Why peak runway pricing is not working." *Journal of Air Transport Management.* Vol. 11. Elsevier Ltd. 2005. p. 417 - 425.

[42] Exemptions were provided for air taxis at JFK and EWR.

[43] Schank. p. 420.

[44] Leo, Flavio. "Boston Logan International Airport's Peak Period Surcharge Regulation - Overview." Power Point Presentation. June 6-8, 2007. Can be found at [http://nextor.org/Conferences/200706_Airport_Access_Rights/Leo.pdf].

[45] Ibid.

[46] Ibid.

[47] Schank. p. 423.

[48] Ibid.

[49] For a further discussion of some of the issues associated with this issue, see Plavin, Z. David. "Using the Revenues from Airport Pricing." Policy Brief No. 68. The Reason Foundation. November 2007. 12 p.

[50] Ball, Michael, Donohue, George, and Hoffman, Karla. Auctions for the Safe, Efficient and Equitable Allocation of Airspace System Resources. Found in Cramton, P., Y. Shoham and R. Steinberg, eds. *Combined Auctions*, MIT Press, Cambridge, Chapter 22, p 507-538. 2006.

[51] Ibid.

[52] Civil Aeronautics Act of 1938, P.L. 75-706, 52 Stat. 973 (1938).

[53] *See id.*

[54] Federal Aviation Act of 1958, P.L. 85-726, 72 Stat. 731 (1958).

[55] *Id.* at § 105 (codified as amended at 49 U.S.C. § 41713(b)(1) (2000)).

[56] 49 U.S.C. § 41713(b)(3) (2000).

[57] Airline Deregulation Act of 1978, P.L. 95-504, 92 Stat 1705 (1978).

[58] *See id.* at § 307 (codified as amended at 49 U.S.C. § 40103(a)(1) (2000) (stating that "[t]he United States Government has exclusive sovereignty of airspace of the United States")).

[59] Department of Transportation Act of 1966, P.L. 89 — 670, § 6(c)(1), 80 Stat. 938 (1966) (codified as amended at 49 U.S.C. § 40103(b)(1) (2000)).

[60] *Hillsborough County v. Automated Med. Lab.*, Inc., 471 U.S. 707, 713 (1985) (quoting Rice v. Santa Fe Elevator Corp., 331 U.S. 218, 230 (1947)).

[61] *Id.* (quoting Rice, 331 U.S. at 230).

[62] *See generally*, Luis G. Zambrano, *Balancing the Rights of Landowners with the Needs of Airports: The Continuing Battle Over Noise*, 66 J. AIR L. & COMM. 445 (2001); see also Ann T. Field & Frances K Davis, *Can the Legal Eagles Use the Ageless Preemption Doctrine to Keep American Aviators Soaring Above the Clouds and into the Twenty-First Century?*, 62 J. AIR L. & COMM. 315 (1997).

[63] Anti-Head Tax Act of 1973, P.L. 93-44 § 7(a), 87 Stat. 90 (1973) (codified as amended at 49 U.S.C. § 40116(b) (2000) (expressly prohibiting "a State, a political subdivision of a State, and any person that has purchased or leased an airport" from levying or collecting "a tax, fee, head charge, or other charge on: (1) an individual traveling in air commerce; (2) the transportation of an individual traveling in air commerce; (3) the sale of air transportation; or (4) the gross receipts from that air commerce or transportation.").

[64] *Id.* (codified as amended at 49 U.S.C. § 40116(d)(2)(A) (2000)).

[65] *See* Airport and Airway Improvement Act of 1982, P.L. 97-248 §§ 503(a)(6), 505(b)(2), 509(b)(1)(E), 511(a)(1)(B), 96 Stat. 673-87 (1982) (codified as amended at 49 U.S.C. § 47107(a)(2)(A)-(B) (2000)).

[66] *Id.* (codified as amended at 49 U.S.C. 47107(a)(13)(A) (stating that "the airport owner or operator will maintain a schedule of charges for use of facilities and services at the airport ... that will make the airport as self-sustaining as possible under the circumstances existing at the airport, including volume of traffic and economy of collection.").

[67] 411 U.S. 624, 626 (1973).

[68] *See id.* at 635 n. 14.

[69] *See American Airlines v. Dep't of Transp.*, 202 F.3d 788, 806 (5th Cir. 2000).

[70] *See Santa Monica Airport Ass'n v. City of Santa Monica*, 659 F.2d 100, 104 (9th Cir. 1981); (limiting the permissible subject matter of local regulations to "aircraft noise and other environmental concerns at the local level"); *see also British Airways v. Port Authorityof New York*, 558 F.2d 75, 84 (2d Cir. 1977) (stating that a proprietor "is vested only with the power to promulgate reasonable, non-arbitrary and non-discriminatory regulations that establish acceptable noise levels for the airport and its immediate environs").

[71] *See National Helicopter Corp. of America*, 137 F.3d 81, 88-89 (2d Cir. 1998).

[72] *See Western Air Lines v. Port Authority of New York and New Jersey*, 658 F. Supp. 925, 958 (S.D.N.Y. 1986).

[73] *See American Airlines*, 202 F.3d at 807.

[74] *Id.* at 808.

[75] *See Aircraft Owners and Pilots Ass'n, v. Port Authority of New York*, 305 F.Supp. 93, 98 (D.C.N.Y. 1968).

[76] *Id.* According to the court, the fee apparently had two exceptions, one for helicopters and the other for air taxis operating pursuant to Port Authority permits. *See id.*

[77] *See id.* at 104.

[78] *See id.* at 107.

[79] *See id.* at 108.

[80] *See id.* at 109.

[81] *See id.* at 109.

[82] *Id.* at 105.

[83] *Id.*

[84] *Id.* at 107-08.

[85] *Id.* at 108.

[86] *Id.* at 109.

[87] 883 F.2d 157 (1st Cir. 1989).

[88] *Id.* at 158.

[89] *Id.* at 159.

[90] Opponents included the National Business Aircraft Association, the Aircraft Owners and Pilots Association and the Regional Airline Association. See id. at 160.

[91] *See id.*

[92] The New England Legal Foundation filed the suit on behalf of itself as well as the Associated Industries of Vermont, Eastern Air Charter, Inc., Montair Flight Service, Inc., Ferns Flying Service, Inc., Emerson Aviation, and two business entities that owned small aircraft using Logan's facilities. *Id.* at 160-61.

[93] The three statutes cited were the Federal Aviation Act of 1958, the Federal Anti-Head Tax Act, and the Airport and Airway Improvement Act of 1982. *Id.* at 161.

[94] U.S. CONST. Art. 1, § 8, cl. 3.

[95] U.S. CONST. Amend. XIV. It should be noted that the Department of Justice, representing the Department of Transportation (DOT), filed an *amicus curiae* brief indicating its interest in the case and suggesting that if the court could not resolve the issues prior to July 1, 1988, that it issue an injunction preventing the implementation of the PACE fee structure pending the outcome of the DOT investigation. *Id.* at 161.

[96] Airline Deregulation Act of 1978, P.L. 95-504 § 105(b)(1), 92 Stat 1705 (1978) (codified as amended at 49 U.S.C. § 41713(b)(3) (2000) (stating that "Nothing in ... this section shall be construed to limit the authority of any State or political subdivision thereof or any interstate agency or other political agency of two or more States as the owner or operator of an airport served by any air carrier certificated by the Board to exercise its proprietary powers and rights.").

[97] *Id.* at 162 (citing *Wardair Canada, Inc. v. Florida Department of Revenue*, 477 U.S. 1 (1986); *Western Airlines, Inc. v. Port Authority of New York and New Jersey*, 817 F.2d 222 (2d Cir. 1987); and *Midway Airlines, Inc. v. County of Westchester*, 584 F.Supp. 436 (S.D.N.Y.1984)

[98] *Id.* (citing *Evansville — Vanderburgh Airport Authority District v. Delta Airlines, Inc.*, 405 U.S. 707, 713, 717 (1972); *American Airlines, Inc. v. Massachusetts Port Authority*, 560 F.2d 1036, 1039 (1st Cir. 1977)).

[99] *Id.* at 162.

[100] The statutes specifically cited were sections 307 and 105 of the Federal Aviation Act of 1958. *Id.*

[101] *Id.* at 163.

[102] *Id.*

[103] *Id.* at 165.

[104] *Id.*

[105] *Id.*

[106] *Id.* at 170 (citing *Chevron U.S.A. v. Natural Resources Defense Council*, 467 U.S. 837, 844 (1984); *McInnis v. Weinberger*, 530 F.2d 55, 63 (1st Cir. 1976).

[107] *Id.* at 170 (citing 49 U.S.C. App. § 1513 (1988) re-codified as amended at 49 U.S.C. § 40116 (2000)).

[108] *Id.* at 171 (citing *Port of Boston Marine Terminal Ass'n v. Rederiaktiebolaget Transatlantic*, 400 U.S. 62, 68 (1970); *see also United States v. Western Pacific R. Co.*, 352 U.S. 59, 63-65 (1956); *Far Eastern Conf. v. United States*, 342 U.S. 570, 573-75 (1952)).

[109] *Id.* at 175 (stating that "the actions of Massport under its new landing fee scheme are even more clearly prohibited by this provision as an interference with air navigation").

[110] *Id.* at 176 (stating that "in keeping with longstanding precedent that requires us to avoid ruling on constitutional issues when non-constitutional grounds are dispositive, we shall not pass upon the questions raised dealing with alleged independent violations of the Interstate Commerce Clause and the Equal Protection and Due Process Clauses.").

[111] Meaning of its own kind/genus or unique in its characteristics. *See* BLACK'S LAW DICTIONARY, 1448 (7th Ed. 1999).

[112] Department of Transportation Act of 1966, P.L. 89 — 670, § 6(c)(1), 80 Stat. 938 (1966) (codified as amended at 49 U.S.C. § 40103(b)(1) (2000)).

[113] *Massport*, 883 F.2d at 165.

[114] "Negative preemption" occurs in cases where the state and/or local authorities are prohibited from regulating even where the responsible federal agency is silent. *See, e.g.,* Susan J. Stabile, *Preemption of State Law by Federal Law: A Question for Congress or the Courts?*, 40 VILL. L. REV. 1, 6-7 (1995) (citing *Arkansas Elec. Coop. Corp. v. Arkansas Pub. Serv. Comm'n*, 461 U.S. 375, 384 (1983); *Ray v. Atlantic Richfield Co.*, 435 U.S. 151, 178 (1978); *Machinists v. Wisconsin Employment Relations Comm'n*, 427 U.S. 132 (1975)).

[115] Independent Offices Appropriation Act of 1952, P.L. 137, § 501, 65 Stat. 268, 290 (1958) (codified as amended at 31 U.S.C. § 9701 (2000)).

[116] *Id.*

[117] *Federal Power Com'n v. New England Power Co.*, 415 U.S. 345, 349 (1974).

[118] *See Air Transport Ass'n of America v. C.A.B.*, 732 F.2d 219, 221 (D.C. Cir. 1984).

[119] 49 U.S.C. § 40116 (2000).

[120] 49 U.S.C. § 47107 (2000).

[121] *Id.* at (a)(2) (emphasis added).

[122] 31 U.S.C. § 3302(b) (2000).

[123] *See, e.g.,* 72 Comp. Gen. 164, 165 — 66 (1993).

[124] *See, e.g., Northern Pipeline Constr. Co. v. Marathon Pipe Line Co.*, 458 U.S. 50 (1982).

[125] *See* 31 U.S.C. § 1341 (2000) (prohibiting the expenditure of federal funds without express authorization by Congress).

In: FAA Airspace Redesign and Congestion Management ISBN: 978-1-60692-710-6
Editor: Sidney R. Rothman

Chapter 3

STATEMENT OF JAMES C. MAY BEFORE THE SUBCOMMITTEE ON AVIATION OF THE COMMITTEE ON TRANSPORTATION AND INFRASTRUCTURE

Air Transport Association

INTRODUCTION

ATA member airlines, which carry more than 90 percent of domestic passenger and cargo traffic, reflect the changing and diverse nature of commercial aviation today. Our membership includes the leading network passenger and low-cost carriers, and both large and small cargo carriers.[1] The significance of this point is that our entire membership is unified in its opposition to the Department of Transportation (DOT)[2] proposed congestion management proposals. As this committee knows, it is not often that our competitive, sometimes fractious, membership fully aligns on a policy issue. Congestion management is one of those issues.

ATA is also aligned with the airport community, including the Port Authority of New York and New Jersey, in opposing the DOT auction proposal. Perhaps an even rarer state of affairs.

DOT proposals to "manage" congestion in New York airspace – slot auctions and congestion pricing – reflect what is manifestly a very poor policy judgment about how to address delays in the New York area. It is a poor policy choice because these proposals are designed to conceal the root problem underlying delays.

Congestion and delays in New York result from several factors, but the primary, driving factor is DOT failure to supply the airspace and air traffic management infrastructure this *country* needs. The commerce of the United States – indeed of the world – drives airline scheduling and by seeking to curb artificially the demand for airspace and air traffic services, DOT proposals harm U.S. commerce and the national economy, harm the local economy of New York City, and harm the competitiveness of U.S. airlines in the global aviation marketplace. Simply put, these proposals are a confession of failure.

Equally important, DOT lacks statutory authority for its congestion management proposals. The fees associated with the proposed auctions are new user fees that Congress has prohibited.[3] Furthermore, when Congress wants to grant an agency authority to conduct auctions, it knows how to do so. It has not done so here. Indeed, DOT has acknowledged, on more than one occasion, that it does not have authority to mandate congestion management measures.[4] This leads inexorably to the question of why does DOT continue to waste valuable taxpayer dollars pursing unlawful measures that will not work?

The real solution to congestion lies in DOT pushing ahead with the tools it does have. Instead of trying to manipulate airline scheduling through artificial means, DOT should manage the airspace and the air traffic control system more effectively and efficiently: implement airspace redesign and related initiatives; work with the Department of Defense and Congress, as necessary, to open up new airways on a permanent basis; accelerate development and implementation of the technologies that will bring us NextGen; and work with the Port Authority of New York and New Jersey and Philadelphia International Airport, and the airlines, to implement the numerous near-term capacity enhancement measures that were identified by the New York Aviation Rulemaking Committee last year. To the extent operations at New York (or other) airports must be capped, DOT should adopt fully the *Worldwide Scheduling Guidelines*, which contains a well-established and accepted slot allocation process, coupled with a vibrant and transparent secondary slot market.

Finally, DOT has lost sight of the ancient maxim, *primum non nocere*, "first, do no harm." The U.S. airline industry is reeling from oil shock like no other U.S. industry and it is unlikely to survive in its present form. Eight U.S. airlines have ceased operating since the end of 2007 and another airline is operating in Chapter 11. Instead of experimenting with illegal and ill-conceived plans to suppress demand for air transportation services, the Department should assess what it can do to relieve the industry of unnecessary costs and regulatory burdens.

DOT Congestion Management Proposals Ignore Reality

DOT Congestion Management Proposals

DOT has proposed two measures to "manage" congestion in the New York region. The first proposal would modify the joint FAA/DOT formal policy on airport rates and charges to permit – indeed encourage – airports to increase the costs they charge to airlines for operating during congested time periods. Because airport charges must, as a matter of law, be cost-based, *DOT has proposed measures that would allow airports to artificially increase the costs that can be passed on to airlines.*

The second proposal is an experiment to auction slots at each of the three primary New York City airports. Each year for the next five years, FAA will confiscate slots from carriers and then auction them off to the highest bidder.[5] In ten years, all slots would automatically terminate and revert back to the FAA, leaving carriers without any idea of their ability to operate their schedules. This feature is inconsistent with encouraging carriers to invest in the operations, facilities, aircraft and employees necessary to compete at LGA.

As discussed in Section III below, both of these proposals are legally deficient and, for that reason, cannot be implemented. However, they are also the outcome of poor policy judgments because they ignore operational reality.

Weather, Not Air Carrier Schedules, Causes Delay

In the FAA's own words:

> Bad weather causes 70 percent of all delays. The situation is worse during the summer: unlike winter storms, which take time to develop and move slowly, summer storms can form quickly, stretch for hundreds of miles and travel rapidly over large portions of the country, grounding flights and sending chain-reaction delays throughout the nation's airspace system. *FAA Fact Sheet*, May 22, 2008.[6]

Little can or need be added to this revealing statement. When it comes to airline delays, the chief culprit is weather. It is obvious, of course, that the impact of bad weather is greatest where air traffic is heaviest, such as in the New York region. But, that is not justification for DOT experimental proposals, especially when other contributing factors – discussed below – are considered.

New York Airspace Has Significant Non-Air Carrier Jet Traffic That Contributes to Congestion

We have testified on several occasions that business jet operations are a significant contributor to congestion and delays in the New York region. We pointed out last September, for example, that air carrier and air taxi (primarily regional airline) operations combined accounted for just 53 percent of the New York City activity based on July 2007 data. In the future, this number is likely to decrease further given the schedule reductions airlines have begun to announce.[7]

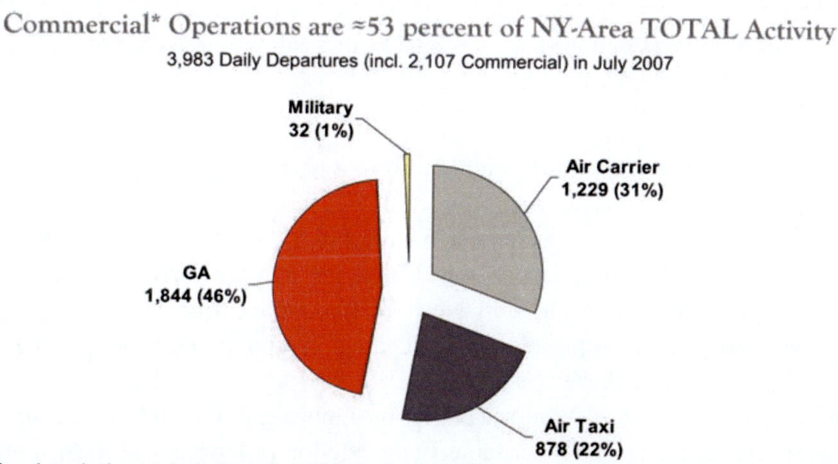

Source: Federal Aviation Administration (FAA) OPSNET * Air Carrier + Air Taxi

Remarkably, this fundamental fact is ignored by DOT demand-management proposals. In light of recent findings by the DOT Inspector General, this oversight is puzzling and should be a serious concern to this Committee.

In March 2008, the DOT Inspector General released a report on use of the National Airspace System("NAS").[8] In his report, the Inspector General found that "business jets' NAS usage is considerable," with "[n]on-air carrier jets accounting for 12 percent of tower and 13 percent of terminal area control services in 2005."[9] The Inspector General added: "To put this in perspective, business jets' tower and terminal area control services in FY 2005 was about one-third of air carrier jets."[10]

Regarding the New York City region, the Inspector General noted:

> the New York TRACON facility handles three large primary airports, [footnote omitted] primarily serving air carriers, and 12 outlying towered airports, primarily serving non-air carriers. *Non-air carriers accounted for 20 percent to 30 percent of the peak level of instrument approach operations at the New York TRACON.* (emphasis added).
>
> [B]oth air carriers and non-air carriers were competing for terminal area control services during the same busy, congested time periods. For example, at the New York TRACON, non-air carriers exhibited the same time of day peaking in demand for terminal services as did air carriers (see figure 5).[11]

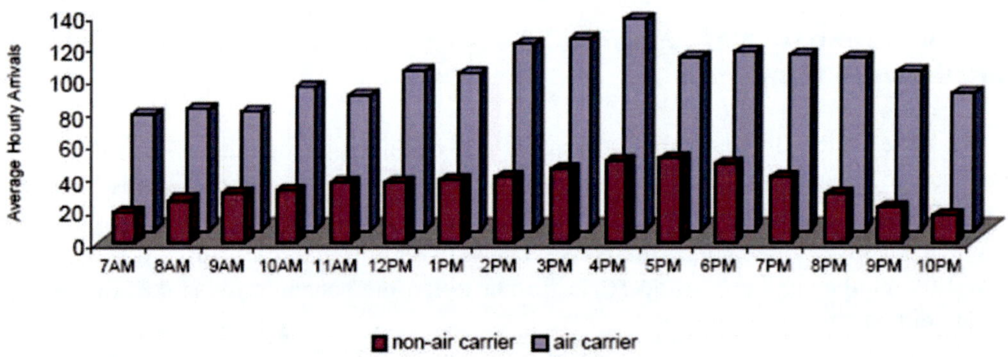

Figure 5. New York Terminal Control Area – FY 2005 Instrument Approach Operations by Hour of Day (Includes Outlying Airports)

Likewise, the Inspector General found that non-air carrier use of the New York Terminal Control Area peaks during the afternoon hours and contributes to congestion.[12]

These findings by the Inspector General further highlight a significant policy deficiency of DOT congestion management proposals: DOT has ignored the very serious and significant contribution to delays and congestion in New York airspace from business jet and high-performance general aviation operations. DOT auction and congestion pricing proposals cannot succeed if they ignore these operations.

The obvious related point, of course, is that by ignoring non-air carrier jet operations in New York airspace, the proposals – particularly the auction proposals – shift the entire cost of reducing congestion and delays (financial and operational) onto air carriers and their passengers and shipping customers. Rather than seeking a fair and balanced solution, DOT's proposals amount to yet another subsidy for business jet owners and other non-air carrier jet operators. Such a burden on commercial air carriers and their customers is patently unfair.

DOT AUCTON AND CONGESTION PRICING PROPOSALS ARE UNLAWFUL

The FAA's own clear and unqualified words in 2006 condemn its current auction and congestion pricing proposals:

> [A] legislative proposal to Congress . . . will seek authority to utilize market-based mechanisms at LaGuardia in the future. Such legislation would be necessary to employ market-based approaches such as auctions or congestion pricing at LaGuardia because the FAA currently does not have the statutory authority to assess market-clearing charges for a landing or departure authorization. If Congress approves the use of market-based mechanisms as we plan to propose, a new rulemaking would be necessary to implement such measures at LaGuardia. FAA Notice of Proposed Rulemaking – *Congestion Management Rule for LaGuardia Airport*, 71 Fed. Reg. 51,360, 51,362 (Aug. 29, 2006).

FAA could not have been more direct: We do not have the authority to impose so-called "market-based approaches," so we will ask Congress to give us that authority. True to its word, the administration's FAA reauthorization bill included provisions to give FAA authority to implement market mechanisms to allocate slots. But, as we all know, that bill has not passed yet. Consequently, by its own admission, FAA lacks statutory authority for the auction and congestion pricing rulemakings.

That should be the end of the story, but it is not. FAA now attempts to qualify its 2006 admission by stating that it analyzed the issue of statutory authority too narrowly. While still acknowledging that it lacks *regulatory* authority to impose market-based mechanisms because of the annual appropriations prohibition against promulgating or collecting new user fees,[13] it conveniently claims it has now determined that it can work around that limitation by exercising *procurement and transaction authority*[14] to auction slots at the three primary New York area airports. That view of the law is incorrect for several reasons, which are discussed at length in ATA comments in response to the FAA supplemental notice of proposed rulemaking (the "SNPRM").[15]

Briefly, the FAA does not have authority to auction slots using its procurement and transaction authority because:

- *It cannot do indirectly what it is prohibited from doing directly. The "transaction authority" that the FAA argues is independent of its regulatory authority (and which it cites as the statutory basis for auctioning and leasing slots) is, in fact, linked to its regulatory authority because the FAA may engage in transactions only if they are "necessary to carry out the functions of the Administrator and the Administration." Thus, in this instance, the "function of the Administrator" being exercised is the function of managing the navigable airspace, which is a regulatory function required by 49 U.S.C. § 40103(b) ("The*

Administrator shall...*assign by regulation or order the use of the airspace..."). But, as even the FAA acknowledges, the appropriations prohibition on new user fees discussed previously prevents it from using its regulatory authority for market mechanisms. Therefore, FAA cannot use its* transaction *authority to implement a prohibited* regulatory *action.*

- *49 U.S.C. § 40103(b), on which FAA relies to promulgate the auction rules, provides specific but limited authority for the FAA to regulate the airspace to ensure the safety of aircraft and the efficient use of airspace and to prescribe* air traffic regulations *on the flight of aircraft. This language does not remotely suggest that the FAA is authorized to auction off the right to conduct operations in navigable airspace. When Congress wants an agency to conduct such an auction, it is explicit. Congress knows how to authorize auctions when it chooses to do so, as it did in great detail and at great length when it authorized the Federal Communications Commission to conduct spectrum auctions. 47 U.S.C. § 309(j). It has not done so here.*
- *Auctioning slots effectively amounts to the imposition of a tax designed to discourage airlines from using the navigable airspace at congested airports during peak periods. Since only Congress can levy taxes, the FAA cannot impose charges that amount to a tax unless Congress has clearly expressed its intention to delegate such authority to the agency and articulated intelligible guidelines for making the assessments. 49 U.S.C. § 40103(b) cannot be read as constituting such a delegation.*
- *The slots that FAA would create under the SNPRM, which FAA describes as "reservations of airspace," are not considered "property" in the hands of the FAA that the agency can dispose of using its property-management authority under 49 U.S.C. §§ 106(l)(6) & 106(n) and 49 U.S.C. § 40110(a). Those provisions apply to the acquisition and disposition of the FAA's real and personal property. A slot – in essence a license or permission to use navigable airspace – is the product of* regulatory *action by FAA in capping hourly operations at the airport. The resulting permission to use what has become constrained navigable airspace is not real or personal "property" of the FAA – just as other licenses or permits issued by governmental authorities are not "property" of the issuing agency.*[16] *The awkwardness of the FAA proposed lease form for slots underscores the fiction of characterizing slots as property of the FAA that it can dispose of by lease.*
- *Because the auction price for slots would not be cost-based, it would violate the requirements of the Independent Offices Appropriations Act (the "IOAA"), 31 U.S.C. § 9701, which allows agencies to charge recipients of special governmental services for the* cost to the agency *of providing those services. Here, the cost to the FAA would be its actual costs incurred to allocate slots by auction. The slot auction prices that are contemplated under the SNPRM, however, would not be related to those costs as they would result from a bidding process.*
- *The FAA has no authority to determine how auction proceeds would be used. Nothing in the Transportation Code or the IOAA authorizes the FAA to retain the auction proceeds and expend them on "congestion management in the New York City area," as the agency proposes to do under SNPRM Option 1. Instead, the auction proceeds under that option (assuming a market-based auction were otherwise lawful) would have to be deposited into the general fund of the U.S. Treasury pursuant to the Miscellaneous Receipts Statute, 31 U.S.C. § 3302(b). The FAA's expenditure of auction proceeds without a congressional appropriation also could violate the Anti-Deficiency Act, 31 U.S.C. § 1341(a)(1).*

- *Likewise, SNPRM Option 2 (where the original slot holder would be allowed to keep the proceeds net of the FAA's auction-related expenses) is not authorized. Option 2 amounts to the forced sale of slots by unwilling sellers, rather than a transfer from the FAA to a carrier. There is no transfer from the FAA to a carrier, required by the FAA transaction authority construct under 49 U.S.C. §§ 106(l)(6) & 106(n). Nor could the FAA rely on the IOAA as authority for Option 2 because the auction proceeds, which are being retained by the original slot holder, do not constitute a charge made to recoup the cost of special services being provided by the agency.*

Likewise, the proposed rates and charges policy change is illegal. Airports are bound by the principle of "revenue neutrality," which means that the total fees and charges collected from airlines must approximate an airport's cost of providing facilities and services. To the extent the policy change purports to allow airports to violate this principle, it is illegal. Moreover, each of the three individual policy changes DOT proposes is legally deficient. Taken as a whole, the congestion pricing proposal, if adopted, is unlawful and will only inject uncertainty for both airlines and airports, thereby causing controversy and disputes. Such controversies will, in turn, undermine the DOT policy of supporting negotiated agreements because airports will be able to unilaterally impose conditions that otherwise would be subject to negotiation. Airport proprietary powers are limited, and it is our view that airports are preempted by federal law from seeking to affect airline routes and services by means of unilaterally imposed pricing schemes.[17]

Auctions and Congestion Pricing Will Not Reduce Congestion or Delays, and Are Fraught with Problems

At the most fundamental level, auctions and congestion pricing have nothing to do with reducing congestion or delays. They are simply a means of allocating limited airport access. Capping operations, as the FAA has done at the three New York area airports, is the mechanism that reduces delays by limiting the number of operations. This fundamental point cannot be overstated. DOT proposals do not address, and will not impact, congestion or delays.

Auctions

No airport or government agency auctions access to airports. DOT seeks to break new ground. But auctioning airport access is fraught with technical and operational problems and will not work.

The DOT auction proposal requires DOT to design, implement and maintain a slot auction mechanism that accommodates the complexities and interdependencies of airline schedules. Airline schedules at one airport are highly interdependent with schedules and operations at other airports across an airline's system and across the entire day, and limitations imposed by slot holdings at other airports, as well as operating limitations (voluntary curfews, connecting schedules), add layers of complexity. The ability to submit and accept package bids (bids conditioned on winning matched pairs of slots at different airports or sets of slots at the same airport) likely will be a critical factor in an auction system. There is little experience in any context, and none in the airline system context, to serve as a model for developing and operating an efficient auction mechanism that deals with these levels of complexities. It is not clear at all that DOT is capable of developing such a complex auction system on its own or with outside assistance.

Notwithstanding these complexities, DOT anticipates that it will be able to issue a request for proposals, select an auction design vendor, resolve numerous outstanding questions about the auction process and determine the auction design, rules and procedures, obtain and test the auction software, train FAA and carrier personnel, and then implement an auction by December 2008 – in less than six months. DOT aspirations are wholly unrealistic and should be cause for concern.

Congestion Pricing

Congestion pricing has proven to be an utter failure wherever it has been tried. A 2005 survey of the literature addressing congestion pricing along with an analysis of peak pricing schemes in Boston, New York and London concluded that institutional barriers prevent peak pricing from being used effectively in the airport context.[18] At several airports, including Toronto and London Stansted, congestion pricing programs (perhaps better described as peak-hour charges) simply have been ineffective in reducing demand for airport access. Congestion during peak hours remains a problem and there is a waiting list for access and/or more slots.

At other locations, including the Cayman Islands and Haiti, so-called congestion charges are merely excess charges to raise revenue. Although characterized as congestion charges, demand at these airports does not exceed capacity and there is no congestion problem.

From a policy perspective, congestion pricing will not work in the airline context because, unlike toll roads where commuters have a choice of routes (secondary surface roads, primary surface arteries or interstate highways), typically no choice exists for airlines and their customers. Passengers will continue to demand flights at particular times and in particular markets, and airlines will respond with schedules to meet that demand.

Also, unlike the roadway and variably priced electricity examples, where the driver or electricity customer pay the fee directly, congestion fees would be imposed on airlines and not the consumer who drives airline scheduling. This means the ultimate consumer is shielded, either partially or completely, from the congestion fee, thereby making it ineffectual at changing consumer behavior.

Finally, congestion fees ignore the investment by airlines in routes, equipment, facilities and personnel. Those investments in most cases have been substantial and, for this reason, airlines will be unwilling to alter their service patterns in response to congestion fees. Airlines will be forced to try to pass them on to consumers, which is difficult in the industry's highly competitive environment, or to simply absorb them, an alternative that cannot be sustained in today's cost environment.

FAA HAS THE TOOLS: AIRSPACE REDESIGN, NEXTGEN AND THE ARC CAPACITY ENHANCING MEASURES

As we have said before, the solution to delays lies not in suppressing demand, but in expanding capacity to satisfy that demand, thereby fostering the health of not just the airline industry, but the entire U.S. economy. All stakeholders, but especially FAA, must be relentless in their efforts to enhance capacity. It is well documented that delays in New York impact the entire National Airspace System. Accordingly, FAA should devote whatever resources are necessary to enhance capacity and operational efficiency in the New York area.

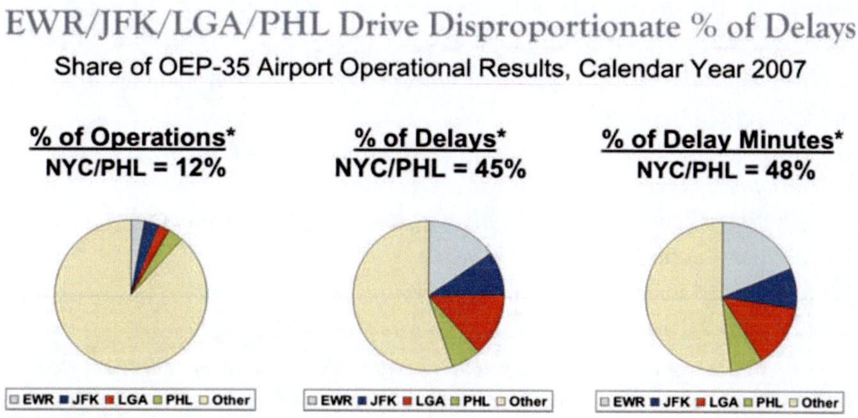

* Share of OEP 35 total
Source: ATA analysis of FAA OPSNET

One measure that we recommended last September, and that we continue to press, is accelerating implementation of the New York/New Jersey/Philadelphia airspace redesign project. This project will enhance both capacity and the efficiency of operations in the Northeast, while reducing the overall number of people exposed to aircraft noise. It is a win-win. Although FAA is moving forward on this project, more can be done to accelerate its implementation, and this committee can assist that effort by keeping the FAA focused on achieving results.

Another measure that we recommended was the appointment of a single person to be responsible for managing and implementing all of the capacity enhancement measures in the New York region – the "New York Czar." That person was appointed recently (although at a much lower reporting level than we had anticipated) and we look forward to working closely with her to affect positive change in New York. This committee can assist this process by holding the FAA accountable for establishing clear objectives and metrics to measure performance to those objectives, and by requiring regular reports.

Getting to NextGen, the FAA Next Generation Air Traffic Management System, is of critical importance. It is a massive undertaking with many moving parts, and FAA must avoid the failures of past large-scale development and acquisition projects. All of the agencies that have a role in this effort but, particularly the FAA as lead agency, must stay focused and devote the resources necessary to get it done as quickly as possible. Again, oversight by this committee will assist this effort.

Last fall, ATA participated on the DOT New York Aviation Rulemaking Committee, and was instrumental in helping to develop a list of 77 delay reduction initiatives for the New York area airports. FAA committed to completing 17 of those initiatives by this summer. As of this date, 12 have been completed and five more are in process; an additional 21 measures also are in process. We and our members will work with the New York Czar to assist these efforts and push for their early completion, and we look forward to working with this committee to assure the success of these efforts.

In addition to the critical measures noted above, ATA has urged the FAA to move forward as quickly as possible on the following recommendations:

- *Ensure real-time access to military airspace*
- *Repair its relationship with the controller workforce*
- *Increase the controller workforce at any/all New York area facilities in order to achieve maximum operational efficiency*
- *Take advantage of new RNP routes in the West Atlantic and leverage new automation tools that reroute flights around weather*
- *Use Airspace Flow Programs to filter business jets out of congested chokepoints during peak periods – especially afternoon peaks*
- *Utilize multiple runways, including converging runway operations where appropriate*
- *Assign scheduled operations a higher priority than other system users*
- *Improve surface management systems (traffic flows between runways and gates) – in particular continue accelerated deployment of ASDE-X*
- *Eliminate miles-in-trail departure restrictions to airports greater than 500 miles away*
- *Expanded use of low-altitude arrival and departure routes ("capping" and "tunneling")*
- *Realign/relocate arrival, departure and overflight routes to avoid conflicts that drive inefficient routings*

Now, More Than Ever, Do No Harm

When ATA testified last September, we were optimistic that the industry finally had turned the corner from the deep, post-9/11 downturn. We stated then:

> It is safe to say that the U.S. airline industry is in a recovery period from the extreme downturn experienced between 2001 and 2005, when the industry sustained over $35 billion in net losses. In 2006 the industry earned $3 billion net profit, and we project a $5 billion net profit for 2007. Airline employment is on the rise, as is capital spending, which is good news for airlines and their shareholders, employees and the many local economies that depend on a healthy airline industry to drive commercial activity, jobs and tourism.

Unfortunately, our optimism for a continued recovery has been crushed under the weight of skyrocketing fuel prices that could not have been predicted. Today, just nine months later, the U.S. airline industry faces losses that will rival, if not exceed, the losses from the 2001-2005 period. The meteoric rise in jet fuel prices – to prices never imagined in anyone's worst nightmare business case – is driving the industry to the brink of imploding. The numbers tell the story:

- *Crude oil hit an all-time high of $138.51 per barrel on Friday, June 6, 2008, marking a one-day rise of $10.58 – an increase of over 8 percent.*
- *Several analysts predict that oil will hit $150 per barrel in July, and some are predicting even higher prices by year-end.*
- *From January 1, 2008, through June 6, 2008:*
    - Crude oil spot prices averaged $107.52 per barrel, versus $60.55 during the same period in 2007
    - Jet fuel prices averaged $134.18 per barrel, versus $80.37 during the same period in 2007
    - Gasoline prices averaged $114.40 per barrel, versus $83.03 during the same period in 2007
    - Jet fuel prices averaged $26.66 per barrel more than crude oil and $19.78 per barrel more than gasoline
- The portion of an airline ticket needed to pay for fuel is now 40 percent, *up from 15 percent in 2000*
- *Scheduled air service has been eliminated in almost 60 communities nationwide that had service in 2007*
    - Nearly 40 additional communities are slated to lose all scheduled air service later in 2008
- *Mainline domestic capacity is being slashed. Some examples:*
    - Continental will reduce domestic mainline capacity by 11 percent in the fourth quarter
    - US Airways will reduce domestic mainline capacity by 6 to 8 percent in the fourth quarter 2008, followed by an additional 7 to 9 percent in 2009
    - United will reduce domestic mainline capacity by 14 percent in the fourth quarter 2008, with additional reductions planned for 2009
    - American will reduce domestic mainline capacity by 11 to 12 percent in the fourth quarter 2008
- *The industry is rapidly approaching 20,000 job cuts and early-out offers*
- *Eight U.S. airlines have gone out of business since the end of 2007 and another is operating in bankruptcy*
- *U.S. airlines are projected to spend $61 billion on fuel this year, $20 billion more than in 2007 – an increase equivalent to the compensation and benefits of 267,000 airline workers or the acquisition of 286 new jets*

- *ATA forecasts a full-year industry loss of $5 to $10 billion for 2008, and absent a dramatic drop in the price of fuel, a multi-billion dollar loss again in 2009. JP Morgan projects an operating loss of $7.2 billion in 2008 and an even higher operating loss in 2009*

Under these circumstances, the government's first reaction should be "do no harm" – avoid adding unnecessary costs and regulatory burdens. Simply put, the industry needs the government to apply a degree of critical analysis to its own actions to determine what needs to be done to ensure public safety and security, what needs to be done to improve operations and efficiency, and then take only those actions that are necessary at this time. Experimenting with demand management in New York is not needed now, regardless of ones views on its merits. Now is not the time to conduct an experiment that will add out-ofpocket expense for airlines, and create greater uncertainty about schedules and aircraft utilization.

Furthermore, in light of dramatic schedule changes and reductions that airlines have announced and which become effective later this year, it is likely that the New York region will see a noticeable improvement in delays. Although the schedule changes will not impact the level of delays this summer, it is likely that delays will be positively impacted next summer. This is another reason DOT should not advance its proposals now.

CONCLUSION

Instead of advancing illegal and ill-conceived notions intended to suppress demand – at a time when airlines are attempting to survive (literally) unheard of fuel prices coupled with consumer demand that is falling through the floor because of a weak U.S. economy – DOT and the FAA should be focusing on fixing the underlying problem: insufficient airspace capacity in the New York region and an aged and inadequate air traffic control system. Stifling demand will have serious adverse consequences for airlines, consumers and New York area residents who rely on passenger and cargo air transportation services and a vibrant tourism industry.

END NOTES

[1] ATA is the principal trade and service organization of the U.S. scheduled airline industry. The members of the association are: ABX Air, Inc.; AirTran Airways; Alaska Airlines, Inc.; American Airlines, Inc.; ASTAR Air Cargo, Inc.; Atlas Air, Inc.; Continental Airlines, Inc.; Delta Air Lines, Inc.; Evergreen International Airlines, Inc.; Federal Express Corporation.; Hawaiian Airlines; JetBlue Airways Corp.; Midwest Airlines; Northwest Airlines, Inc.; Southwest Airlines Co.; United Airlines, Inc.; UPS Airlines; and US Airways, Inc. Associate members are: Air Canada; Air Jamaica; and Mexicana.

[2] References to DOT include the Federal Aviation Administration (FAA).

[3] "[N]one of the funds in this Act shall be available for the Federal Aviation Administration to finalize or implement any regulation that would promulgate new aviation user fees not specifically authorized by law after the date of the enactment of this Act..." 2008 DOT Appropriations Act, Pub.L. 110-161, 121 Stat. 2379, and prior annual appropriations acts.

[4] For example: "In the [2006 LGA] NPRM, the FAA stated that it did not have the authority to reallocate Operating Authorizations via a market-based mechanism...The FAA continues to believe that it cannot rely on a market-based allocation method under a purely regulatory approach, which is why it explicitly sought legislation on this matter." Supplemental Notice of Proposed Rulemaking, 73 Fed. Reg. 20852 (April 17, 2008). See Section III, for further discussion of this point.

[5] DOT proposes two options for how the proceeds will used. Under Option 1, DOT claims the proceeds will be used to mitigate congestion in the New York region; under Option 2, the proceeds would be paid to the air carriers from whom the slots were taken. But under Option 2, more slots would be confiscated – 20 percent instead of 10 percent.

[6] http://www.faa.gov/news/fact_sheets/news_story.cfm?newsId=10227

[7] The multiple factors affecting congestion and delays are addressed at length in our written statement before this Committee on the subject of Airline Delays and Consumer Issues, September 26, 2007, available at http://www.airlines.org/government/testimony/ATA+Testimony+-+Airline+Delays+and+Consumer+Issues.htm.

[8] Report No. CR-2008-28, March 3, 2008.

[9] Id. at 10, 11.

[10] Id. at 12.

[11] Id. at 13-14.

[12] Id. at 13.

[13] See, for example, Pub L. 109-115 and Pub. L. 110-161.

[14] "[T]he FAA's authority is not limited to regulatory action. The agency has independent authority to dispose of property (footnote omitted), and regulatory action is not required prior to the lease of property." 73 Fed. Reg. 20852.

[15] See Comments of Air Transport Association of America, Inc., June 16, 2008, Docket No. FAA-2006-25709.

[16] In contrast to the FAA, which creates the slots by regulatory action but has no property interest in them, airlines to which slots are issued do have a property interest in slots. That property interest is recognized by third parties (including lenders to whom slots may be pledged as collateral) and by the FAA itself (which allows slots to be bought and sold in a secondary market).

[17] "a State . . . or political authority of at least 2 States may not enact or enforce a law, regulation, or other provision having the force and effect of a law related to a price, route, or service of an air carrier." 49 U.S.C. § 41713(b)(1).

[18] Joshua L. Schank, "Solving airside airport congestion: Why peak runway pricing is not working," Journal of Air Transport Management 11 (2005) 417-425.

In: FAA Airspace Redesign and Congestion Management ISBN: 978-1-60692-710-6
Editor: Sidney R. Rothman

Chapter 4

TESTIMONY OF MR. DOUGLAS E. LAVIN COMMITTEE ON TRANSPORTATION AND INFRASTRUCTURE'S SUBCOMMITTEE ON AVIATION UNITED STATES HOUSE OF REPRESENTATIVES

Douglas E. Lavin*

Mr. Chairman, distinguished Members of the Subcommittee.

My name is Douglas Lavin. I am the Regional Vice President for North America for the International Air Transport Association (IATA).

IATA represents 230 carriers engaged in scheduled international transportation of passengers, mail and cargo by air. 71 of those airlines fly into and out of New York area airports on a scheduled basis. All of the major U.S. network carriers are members of IATA.

IATA appreciates the opportunity to brief the Subcommittee on IATA's position on the proposal by the U.S. Department of Transportation's Federal Aviation Administration to manage congestion at New York Area airports. IATA places a high priority on the effective management of these airports because of the direct impact congestion has on our members that fly into and out of the region as well as the secondary impact New York congestion has on operations across the U.S. and around the world.

Since September 2007, DOT and the FAA have focused considerable time and resources in an effort to address congestion in the New York region. Likewise, domestic and international airlines, domestic airports and the associations representing them have made a tremendous investment of time and resources to identify and promote solutions to this challenge on an expedited basis.

Unfortunately, IATA's domestic and foreign member airlines have been uniformly disappointed with the efforts to date by DOT on this important task. Rather than focusing its limited resources on increasing capacity at these gateway airports, DOT, with the support of

* Regional Vice President (North America), International Air Transport Association

the White House, has spent inordinate amounts of time attempting to implement so-called "market mechanisms" to limit carriers' access to New York airports.

A review of DOT activities in this regard over the past 9 months is instructive:

A Short History of DOT's Efforts to Address Congestion in New York

In September 2007, President Bush and Secretary Peters' announced their intention to take steps to address unacceptable congestion at New York airports. In the short term, DOT designated Newark and JFK as congested airports, and subsequently implemented a temporary cap and the allocation of slots roughly following the principles of the Worldwide Scheduling Guidelines (WSG), an internationally recognized set of standards to manage scarce airport resources. They also worked with the Department of Defense to open restricted airspace during peak holiday periods. Finally, they created the Aviation Rulemaking Committee (ARC), made up of DOT officials and representatives from domestic and foreign airlines and the New York Port Authority, to advise Secretary Peters on how to address NY congestion.

From September to December 2007, the ARC met on a weekly basis to identify short and longer-term solutions to congestion at New York area airports. Five working groups were established to study and report back to the Secretary on such options as operational and infrastructure improvements, the use of socalled "market mechanisms" (e.g. auctions/ congestion pricing), gate utilization, priority air traffic preferences and the WSG and other "administrative options." In the end, the large majority of the airlines participating and the Port Authority made it clear to Secretary Peters that:

- *FAA and DOT need to place a high priority on increasing capacity at New York airports through the expedited implementation of NextGen technologies and procedures as well as the New York Airspace Redesign;*
- *Congestion pricing and auctions are not effective means of either addressing congestion or allocating scare resources; and*
- *While operational and infrastructure improvements are being made, DOT should manage congestion within the general framework of the WSG*

Despite this clear and unequivocal guidance, DOT announced in January 2008 that it intended to lease any new capacity that came on line at JFK via an auction process. It announced an identical approach for Newark in March 2008. Equally surprising given the ARC consensus against congestion pricing, DOT announced in January 2008 proposed amendments to the DOT 1996 Policy Regarding Rates and Charges that would in effect give airports the ability to use pricing to manage congestion under very broadly defined circumstances. DOT acknowledged that this was their best opportunity to impose congestion pricing without having to go to the Congress for authority to in effect impose taxes.[1] IATA strongly protested this action as a contravention of U.S. law, ICAO policy and international practice that airport charges must be revenue neutral, cost based and tied to services actually provided to airlines. The industry is now waiting for what we fear will be a final DOT rule implementing this illegal pricing scheme.

Finally, on April 17, 2008, DOT issued a Supplemental Notice of Proposed Rulemaking (SNRPM) extending caps at LaGuardia through 2019, followed by a May 16, 2008 Notice of Proposed Rulemaking (NPRM) doing the same at Newark and JFK. Rather than implementing the WSG to manage these airports, DOT argued that U.S. sovereignty over and control of the airspace gives it a property interest in the slots at congested airports. DOT proposes to lease the slots at these airports for a period of ten years. More troubling, DOT proposes to withdraw a significant number of slots from airlines operating at the three airports and auction them off to the highest bidder as a means to increase competition. IATA considers this yet another example of DOT seeking to conduct economics experiments with New York at a time when its focus should be on managing existing congestion and undertaking urgent efforts to expand capacity at these three airports.

Discussion

Since September 2007, DOT has set forth three overall goals for addressing congestion in New York: reduce congestion, allocate scarce airport resources and minimize disruption associated with implementing any suggested improvements.[2] It is useful to review DOT's progress in regard to all three goals since September 2007:

1. Reduce Congestion

The most effective step DOT has taken over the past nine months to address congestion in New York is to designate Newark and JFK as Level Three airports, which resulted in temporary cap on operations, the designation of an FAA coordinator, and the holding of schedules reduction meetings under the broad terms of the WSG. IATA supported DOT's actions in this regard as long as these measures were accompanied by a renewed commitment to increase capacity at the region's airports. To that end, the ARC made 77 recommendations on ways to increase capacity over the short, medium and long term. The ARC also recommended that DOT appoint a "New York Czar" to oversee rapid deployment of these operational and infrastructure improvements.

IATA has been supportive of the work the FAA has done since September 2007 to increase capacity at Newark, JFK and LaGuardia airports through operational and infrastructure improvements in the region. The ARC identified 27 short-term initiatives to improve delays by implementing available technologies and/or by modernizing outdated policies and procedures. The FAA agreed to address 17 of these initiatives in the short term. While 14 of the original 17 initiatives are complete as of June 1, the remaining three items are delayed due to lack of internal support and likely not to be completed in 2008. Overall, 60% of the 77 ARC recommendations are expected to be completed by the end of 2008. IATA is also pleased with the appointment of Marie Kennington-Gardiner to the position of NY Area Integration Office Director (NY Czar) and hope that the Congress, the FAA and its bureaucracy provides her with the resources and support she needs to address this important task. While much difficult work remains, we believe that the continued focus of the FAA on increasing capacity is the best opportunity to address congestion in New York in the short and long term.

2. Allocation of Scarce Resources

While the FAA has made progress in increasing capacity, DOT has spent the last nine months devising new and untested "market mechanisms" to limit carrier access and, more recently, increase competition at New York airports. The DOT suggests that "simply imposing a cap and then doing nothing to ensure that there are competitive market forces and actual and potential competition is unacceptable."[3]

First, one has to challenge DOT on their assumption that a cap alone would hurt competition at JFK and Newark. JFK was capped from 1969 to 2006 under the High Density Rule (HDR) and Newark has been significantly constrained for some time. Today, the two airports currently have a combined 1,142 daily flights operated by more than 100 U.S. and foreign airlines. 78 airlines at JFK and 58 airlines at Newark operate non-stop flights to 159 and 164 cities around the world respectively. It is anticipated that the majority of these 100 airlines would continue to operate at these airports whether or not a cap is in place. Slot auctions and congestion pricing cannot be justified as needed to increase competition at these airports.

Further, IATA and its member airlines, along with the Port Authority, the ATA and non-IATA members serving the New York market, have consistently argued that DOT should utilize the Worldwide Scheduling Guidelines to manage the limited resources at these key airports. The WSG was established more than 60 years ago specifically to serve as a comprehensive set of guidelines for the "allocation of scarce resources on a fair, transparent and non-discriminatory basis". Today, the WSG is utilized at 140 congested airports worldwide, including at Chicago's O'Hare Airport, to manage the scarce airport resource through the establishment of slots, which permit take-off or landing during specified time periods. Slot allocation under the WSG is based on "historical precedence," wherein a carrier that operates a slot on a consistent basis during the previous season is entitled to assignment of those slots again in the subsequent season. By following this approach, the WSG enables consumers to gain the benefits of regularity of services over time. It also enables airlines to reap the benefits of their investments in developing new routes and/or increased frequencies. Slot allocation is conducted twice a year on a global basis at IATA facilitated meetings under the direction of independent airport coordinators appointed by the respective governments responsible for the particular airports.

DOT officials have argued that the WSG is not an effective way to fairly allocate a scare resource as it rewards airline incumbents at the expense of new entrants. DOT has also argued that implementation of the WSG would in effect result in the incumbent airlines having a property right in the slots at the New York airports, contrary to their belief that the slots are owned by the U.S. Government. While DOT acknowledges that airlines have made multi-million dollar investments to support their operations, they have argued that the "historic precedence" in effect grants airlines valuable government property without compensation.

To the contrary, 60 years of experience around the world has demonstrated that the WSG effectively manages a scarce resource while allowing for meaningful competition. Under the

WSG, carriers who do not operate slots at least 80% of the time during the season for which they were allocated are no longer entitled to historic preference for those slots during the next scheduling season. Those slots not allocated based on historical performance are placed in a "slot pool," 50% of which are set aside for new entrants (defined as carriers holding less than four slots a day at the airport). Further, the WSG establishes several additional criteria that should be considered by the slot coordinator when allocating slots from the slot pool, including encouraging competition, allowing for a mix of operations between short and long haul and consideration of the needs of the traveling public.

During the ARC process, some limited incumbents seeking additional slots at Newark and JFK during congested hours argued that the WSG definition of new entrant, combined with the limited number of slots that become available each year via the "use or lose" provisions make it impossible for new entrants to gain meaningful access to congested airports.

It is interesting to note that DOT dismissed these arguments when denying Virgin America's request that 30 slots be set-aside for them at JFK as part of the cap implementation process. DOT pointed out that even with a cap there is ample opportunity to allow limited incumbents increased access to the congested airport:

> All carriers will have an opportunity to acquire and to retime operations at JFK while this Order is in effect. Under this Order, all carriers have the opportunity to trade with others for Operating Authorizations at times that are more desirable to them. In addition, all U.S. and foreign carriers have the opportunity to lease Operating Authorizations from other carriers for the duration of this Order. [4]

DOT's argument in this regard is reflective of experiences at other congested airports around the world. For example, London Heathrow Airport (LHR) has been operating under EU Slot Regulation since 1993 (based on the WSG). Slot coordination has helped allocate scarce capacity while allowing for increased competition via the WSG "use or lose" provision as well as secondary trading. For example, over the last five years, LHR added 25 new entrants despite this cap, including Continental Airlines, Northwest Airlines, U.S. Airways and Delta, which all gained access via secondary trading following the revised U.S. – EU Air Services Agreement.

Given this experience, we question DOT's belief that the WSG as currently configured does not allow for effective competition at congested airports. However, if that concern remains, it is important to note that there are provisions in the WSG to allow governments to institute "local rules" to address particular concerns that they do not believe are adequately addressed by the WSG principles. For example, DOT has the option of implementing a local rule to foster a secondary market, as called for by DOT in their most recent NPRM. DOT could also expand the definition of a new entrant to allow for greater access for smaller carriers to the slot pool. While implementation of local rules diminishes the strength of the WSG as a global scheduling standard, they are much preferable to governments creating their own unilateral system to allocate congested resources, as is now being proposed by DOT in New York.

3. Minimizing Disruption Associated with Implementing Suggestions

With 74 IATA members flying into and out of the New York Region and accounting for the majority of services offered, we have a significant interest in the implementation of effective management tools at these airports. However, equally importantly, we are concerned about the disruption the DOT proposals could cause to the global aviation system, particularly in the international context. There are four areas of significant concern in this regard.

First, while New York airports are critical in their own right, it is important to recognize that Newark and JFK serve as important hubs in an overall international aviation system. To put it in simple terms, a departure at JFK is an arrival at London or Paris or Beijing and potentially a connection beyond those airports. An efficient global aviation system has as its basis common rules and regulations (such as the WSG) that ensure the seamless operation of hundreds of carriers at thousands of airports across 24 time zones. Throughout DOT's consideration of ways to address congestion at airports, there has been a lack of appreciation for the impact unilateral U.S. "market mechanism" experiments may have on what are by definition complex and interconnected international operations. This includes both the slot confiscation proposed for the New York airports as well as the proposed changes to the DOT rates and charges policy that would serve to encourage airports to manage congestion through pricing mechanisms. DOT's theories to the contrary, airlines may send regional jets during congested hours not to cause delays, but instead to provide feed from local communities to European flights that have only limited departure windows. Using pricing and auctions to change that dynamic will impact large and small communities both here and around the world. Disrupted operations in the name of economics experiments will not serve the public good.

Second, both the congestion pricing and slot confiscation proposals raise the possibility of limiting the growth of international carriers at New York airports. DOT's proposal as drafted would have only a limited short-term impact on foreign carrier operations into New York. However, it is safe to assume that over the 10- year life of this proposed rule, foreign airport operations would be constrained by these provisions. This raises implications for existing bilateral aviation agreements between the United States and other countries where certain access is guaranteed and growth is anticipated.

Third, IATA is concerned about the precedent that the DOT's pursuit of market mechanisms would set for foreign governments. Up until DOT's pricing fixation, airports and governments around the world have been constrained from pursuing congestion pricing and slot auctions because of accepted ICAO and international law that primarily restrict monopoly pricing of airport access to actual costs incurred. The few efforts to establish congestion pricing mechanisms at large airports have been successfully challenged on the basis of discrimination against classes of airlines (including a U.S. Government challenge to proposed congestion pricing at Heathrow). U.S. and foreign carriers have been fortunate that these policies and successful legal challenges have prevented an opening of a "Pandora's Box" of charges and auctions by governments seeking opportunities to raise new revenue from airlines and airports.

Unfortunately, DOT's pursuit of market mechanisms has already served to signal a loosening of accepted international constraints on congestion pricing. For example, the Brazilian aviation authorities (ANAC) recently proposed to increase parking fees at Sao Paulo International Airport on U.S. and other carriers that have long layovers at the airport in an effort to redirect those planes to other airports. This would result in an increase in fees paid by U.S. carriers and others of between $15,000 and $90,000 per flight per day. We do not believe ANAC would have pursued this pricing scheme but for the U.S. Government's signal to the international community that congestion pricing is acceptable. We understand that other governments are now also considering implementing congestion-pricing schemes. Further, the DOT NPRM offers industry some assurances that the funds raised under these pricing schemes would be used for some yet to be defined capacity building projects. While we have doubts about DOT's commitment in this regard, we are very concerned that other governments will simply use funds raised to address general budget shortfalls.

Finally, one cannot minimize the negative impact these provisions would have on the airline industry and its passengers. Slot auctions and congestion pricing will conservatively lead to hundreds of million dollars in additional costs for incumbent airlines. Slot auctions will in effect penalize airlines that have already made significant investments in their operations at these airports by requiring them to pay for the privilege a second time. Airlines are not in a position to accept this additional financial burden. As such, these costs will have to be borne by our passengers in the form of increased ticket prices and fewer flights. To be clear, auctions and congestion pricing will have no impact on delays at New York area airports. Instead, passengers will face further inconvenience, increased cost and decreased service at a time when we can all afford it the least.

CONCLUSION

Airlines are today facing the challenge of remaining viable business entities in a market that has seen a tripling of oil prices since 2006 and a two-fold increase in the last year alone. At currently predicted fuel prices, IATA airlines could face an additional financial burden of $99 billion over the next 12 months compared to 2007. A total of 24 airlines have ceased operations or entered into bankruptcy protection in the past five months. Many more airlines will not survive the New Year. Quite simply, airlines are facing a crises that is beyond that which they experienced post 9/11.

The leaders of the world's airlines met in Istanbul on June 2-3, 2008 as part of the IATA World Aviation Summit. The CEOs of these airlines adopted a resolution at that Summit calling on governments and the entire industry value chain to show leadership and responsibility in this time of crises. They called on governments to refrain from imposing multiple and additional punitive taxes and charges that will only deepen this crises while at the same time take the steps necessary to modernize the air transport infrastructure eliminate wasteful fuel consumption and emissions.

DOT's determination to significantly raise costs to airlines through "market mechanism" economics experiments is exactly the type of activity this industry cannot afford during this crisis. It is estimated that international aviation supports 7% of global GDP and 32 million jobs. Proposals like these coming out of DOT cannot help but having a negative impact on this important economic engine. We therefore urge the U.S. Congress to direct DOT to:

- *Continue to focus all available resources on making needed operational and infrastructure improvements in the New York region to increase capacity at these airports*

- *Include Newark, JFK and LaGuardia in the WSG process until such time as new capacity measures make the cap unnecessary*
- *Cease all efforts to experiment with "market mechanisms"*

Thank you for your consideration.

END NOTES

[1] Under Article 1 of the Constitution, Congress is the "sole organ for levying taxes." National Cable Television Association v. United States, 415 U.S. 336, 340 (1974). The Supreme Court has recognized that assessments intended to " discourage [an] activity . . . are in the nature of 'taxes' which under our constitutional regime are traditionally levied by Congress." Id. at 341. Congestion pricing are classic taxes because these measures are designed to discourage takeoffs and landings during peak hours by making operations during those times more expensive. Congress may delegate its taxing authority but it must be "clearly delineated."

[2] NY ARC Report, December 13, 2007, page 1

[3] 73 Fed Reg. 29630 (May 21, 2008).

[4] 73 Fed Reg. 3514 (January 18, 2008)

In: FAA Airspace Redesign and Congestion Management ISBN: 978-1-60692-710-6
Editor: Sidney R. Rothman

Chapter 5

TESTIMONY OF WILLIAM R. DECOTA BEFORE THE UNITED STATES HOUSE OF REPRESENTATIVES, COMMITTEE ON TRANSPORTATION AND INFRASTRUCTURE, SUBCOMMITTEE ON AVIATION REGARDING CONGESTION MANAGEMENT IN THE NEW YORK AIRSPACE

William R. Decota*

Good morning. I am William R. DeCota, the Director of Aviation for the Port Authority of NY & NJ, which operates LaGuardia, John F. Kennedy International, Newark Liberty International, Stewart International and Teterboro airports.

The Port Authority is committed to promoting safe and efficient air travel connecting the New York metropolitan region with the rest of the world. Our three major airports handle over 109 million passengers and 2.7 million tons of cargo and generate over $57 billion in annual economic activity. The Port Authority has taken various steps to enhance capacity and meet the demand for air travel. Over the past ten years, we have invested over $15 billion in upgrades to our airports and have programmed an additional $6 billion over the next decade. In November 2007, we purchased the leasehold interest in Stewart International to help meet the growing regional demand for air travel. Also in 2007, the Port Authority's Chairman, Anthony Coscia, convened the Flight Delay Task Force, composed of senior airline, business, FAA and state and local leaders which identified over 77 initiatives to improve capacity and reduce delays at our airports.

We also know that the FAA has made investments in aeronautical systems. However, despite these investments, the FAA has been unable to accommodate the same number of

* Aviation Director, The Port Authority of New York & New Jersey

annual operations in the New York region as they were a decade ago. We are confounded by this.

The Port Authority has also attempted to work with U.S. DOT and FAA officials on the problem of flight delays for nearly a decade.

In April 2000, Congress had the noble goal of opening up access to the nation's most important, popular, yet constrained airports to stimulate competition and provide opportunity for entry to new markets. For the New York area airports, Congress allowed the FAA and DOT nearly seven years to figure out an appropriate framework.

The FAA took the challenge, engaging the National Center of Excellence for Aviation Research (NEXTOR) universities to examine a range of administrative as well as market-based solutions. No one can deny that the FAA did not spend a lot of time and money on this research effort.

However, we grew to fear these academic exercises. While the professors might be able to explain "in theory" how a market-based solution could solve all the problems at my airports, they were never able to demonstrate how it would work in practice. As the operator of the New York area's key airports, with our proprietary rights and responsibilities to allocate gates and terminal facilities and manage the groundside facilities, the Port Authority attempted to fully engage the DOT and FAA as well as the airlines and other key stakeholders in evaluating potential options to enhance capacity and efficiency and to promote competition and diversity of service at the airports.

Unfortunately, after seven years we are no closer to a workable solution. Rather, the Administration has chosen to impose an approach that we, as the airport operator, think is not only illegal but also disastrous; that the vast majority of the carriers, including the legacy and low cost carriers, are opposed to; and that consumers, represented by such groups as the Air Travelers Association and the Business Travelers Coalition, believe will ultimately harm rather than help them.

The SNPRM for La Guardia and the NPRMs for JFK and Newark are attempting to impose slot auctions despite our collective concerns. We particularly regret that the LGA SNPRM provided only 60 days for comment despite being incomplete; it fails to provide either specific auction rules or the statutory authority for such activity by the FAA. It is also deficient in its supporting data and analyses. Despite requests from both airport and airline stakeholders for a reasonable comment period and corrected information necessary for a proper evaluation of the proposals, it appears that the priority is to fast-track these halfbaked, unpopular proposals for implementation before the end of the year.

We are concerned that the Administration's approach is fundamentally flawed, unworkable, unresponsive to the problem, and disruptive to the airport, to air carriers schedules, and ultimately, to the traveling public.

To start, we believe that the FAA lacks the statutory authority to lease slots at rates determined by an auction. *Nowhere in the Federal Aviation Act is there a declaration that slots are property* and *nowhere is there the authorization to auction or lease slots for monetary compensation there from*. We believe that the FAA would need Congressional authority to lease slots at rates determined by auction. The FCC's auction of radio wave spectrum serves as an example of Congressional authorization required to endow an agency with the power to raise revenue by requiring payment in return for permitting someone to engage in regulated conduct. The FCC, like the FAA, did not have the authority to raise revenue by permitting regulated conduct (there broadcasting, here flying into LaGuardia) until Congress granted it that authority in 1993 to auction off permission to use specified frequencies. We believe that without Congressional authorization, the FAA's auctioning of "slots" amounts to an unauthorized tax without the Congressional authorization to do so.

We are also very concerned that these proposals represent an attack on the proprietary rights of airport operators and will do significant damage to longstanding business arrangements at the New York area airports. However, rather than address this issue at this time, I want to focus my comments on the shortcomings of DOT's scheme.

The SNPRM fails to achieve one of its two primary objectives – reducing congestion and delays at LaGuardia. Despite no changes in airport capacity, and declining aircraft activity, LaGuardia delays have increased by more than 50% from 2004 to 2008. The Port Authority therefore concludes that airspace capacity has declined. The FAA SNPRM clearly defines airspace management as the FAA's responsibility. The SNPRM fails to acknowledge that airspace capacity has declined, fails to provide any evidence or analysis of what the new airspace capacity is, fails to provide any indication of whether the FAA can restore airspace capacity to previous levels, and fails to establish the cap at a level that matches current airspace capacity which is now below runway capacity. While the DOT and NEXTOR have been obsessed with promoting auction schemes, they have failed to address one of their core responsibilities.

Auctions Are *Not* the Solution

For seven years the NEXTOR professors tried to show us how auctions could work in the aviation context. With great theoretical detail, they spoke on the mechanics of the auction and their philosophical belief that a market-based solution would solve all issues. However, each of the NEXTOR presentations would stumble at the same place – there was no recognition of the real world concern of how the winning bidder would have access to the appropriate gate, check-in facilities, baggage belts, lounges, and other terminal and landside facilities needed to actually operate the slot.

A member of my staff referred to this as the "Fatima moment". After espousing all wonders of how an auction would work, the professors would simple say that the Port Authority would figure out how to provide operational sustenance to the new carrier – that is, a miracle would occur.

From this airport operator's perspective, it is not that easy, and I cannot operate my airports in this manner.

We have discussed these concerns at length with the FAA, DOT and their consultants. We gave them numerous tours of the Airport and identified the various constraints that we operate under at LaGuardia. Despite these efforts, the Administration's original NPRM for LaGuardia contained some rather disturbing language on this subject. Although they acknowledged the landside constraints and our existing business arrangements, they and NEXTOR tried to assume them away by reminding the Port Authority about its grant assurances that ensure access to its airports. While we have made and continue to make reasonable efforts to accommodate new entrants and others at our airports, we cannot accept wholesale disruption of our operations.

Their proposal is fundamentally flawed in that it did not properly marry the airside capacity with landside capacity. We can only imagine what would have ensued if DOT and their professors were around when the transcontinental railroad was constructed in the 1860's – the east and the west would have never met in Promontory Summit, Utah.

Gates and Terminal Facilities Are Not Interchangeable

We applaud this Committee's recognition that airport gates are not interchangeable in hearings conducted in 2000. However, the experts at DOT fail to acknowledge this fact. A wide-body aircraft cannot be accommodated at a regional jet gate. A new entrant carrier cannot operate efficiently if its flights can only be accommodated in different terminals. Furthermore, the Port Authority has developed various business arrangements – legal, binding contracts -- with our airline tenants that provide the operational framework and financial underpinnings for the development of the airports. These contracts cannot be ignored.

The professors providing auction advice to the Administration admitted in a publication released last year that they "… ***assume*** that the FAA has established the number of slots that will be available for auction based on runway and landside capacities. The leaseholder of that slot is given the right to trade or sell this slot for any portion of the leasing period in a secondary market. The leaseholder also receives the corresponding rights to terminal space, e.g. ticketing, baggage, and gating facilities and will pay the "going rate" for these facilities based on the current long-term contracts with the local airport authority."

This is not the case. And, it highlights the fundamental flaw in the Administration's scheme. Despite some recent efforts by the FAA to address flight delays and congestion and identify options to enhance capacity, the DOT has been obsessed with imposing so-called "market-based" auctions as an ideological solution ignoring all practical realities and shortcomings.

And, we believe that these shortcomings are significant.

We are concerned that auctions will reduce, not enhance, competition. The auction schemes will strongly favor those larger airlines with the most cash on hand to buy slots at a price above those with less cash on hand. It is possible that the larger carriers could use their relatively stronger balance sheets (despite the current financial challenges faced by the industry) to outbid the smaller, nonlegacy airlines that help stimulate competition. The "value" of a slot could be driven by the benefits a large carrier would reap from eliminating competitors rather than from the operation of the slot itself. The SNPRM appears to encourage this behavior since it explicitly exempts so-called Unrestricted Slots (which are slots awarded via an auction) from any minimum usage standard. In fact, our analysis indicates that the low-cost carriers that have gained access to LaGuardia will lose the largest share of their operations. The result could be a significant increase in airfares.

We are also concerned that the number of destinations, particularly small communities, served from our airports will be impacted. The Port Authority has long been an advocate for ensuring that service to small communities is preserved and our own analyses has shown that this service was particularly vulnerable to market-based solutions. We have suggested various proposals over the years to include either a set-aside or slots or gates for small communities.

The impact on small community service has been evident during the first NEXTOR exercises conducted in 2004 about LGA with airline and airport participants. The results showed that communities with populations under 500,000 were likely to lose frequency of service, if not lose the service completely. In a presentation prepared for the Transportation

Research Board in January 2005, NEXTOR acknowledged that for communities of this size, the number of frequencies was slashed from 130 per day to 60 and for communities between 500,000 and 1 million, the number of operations plummeted from 118 per day to 60.

Let me give you an idea of the service that is vulnerable: cities such as, Huntsville, Lexington, Des Moines, Flint, Bangor, Madison, Ithaca, Roanoke, Knoxville... unfortunately I could go on ... could not only lose service to LGA, but to the greater New York City area. Others, such as Myrtle Beach, Columbus, Richmond, Savannah, Jacksonville, and Buffalo could see reduced service and higher air fares.

And this is not only a New York phenomenon. In 2005 NEXTOR conducted a study called: Using Auction-Based Slot Allocation for Traffic Demand Management at Hartsfield Atlanta International Airport: A Case Study. It found that "As expected, when money is the only determination factor, small aircraft with less revenue are the first to be eliminated from the auction process.... As a result, potential distortions of the marketplace could impact negatively access to small communities or fair competition."

More recent analysis from the NEXTOR professors has identified three communities that would lose service altogether, but given the economics of the aviation industry at the moment; I firmly believe that this analysis is woefully underestimating the impact.

Small communities in general are served by fewer carriers than larger ones. There are 21 small and non-hub markets with service to LaGuardia on a single carrier. When a small community loses service on its only carrier to LaGuardia,it is unlikely that community will regain the service, since the carrier gaining the slots in the auction may not even serve the community in question. The FAA did acknowledge that the cost per seat of the auction price is "2 to 6 times higher for regional jets and turboprops". Because that most of the small and non-hub markets are served by regional jets and turboprop aircraft, the auction schemes give these markets a significant cost disadvantage against larger markets. There is a very real risk that a significant portion of the small communities will lose their air service to LaGuardia and not gain a replacement carrier.

It is clear to us, given these fundamental flaws, that DOT's proposed auction scheme is not the right solution for the New York airports. Although market-based solutions may have a place in the aviation industry, such approaches must be carefully crafted together with the airport operator who is in the best position to know how and when it can be employed.

We firmly believe that there are proven alternatives, such as an IATA-like scheduling process in conjunction with both local leasing practices and policies and a specific federal set-aside for small community service. If adopted, this approach would maintain competition, diversity of service and reasonable air fares for the traveling public. An IATA-like process would incorporate the airside and landside capacities into the slot allocation process as well as providing accommodations for new entrants. Local leasing policies would be important to promote the efficient use of facilities consistent with market demand. Federal rules could identify a certain portion of slots exclusively for service to small and non-hub communities.

We also believe that transparent buy/sell rules, along with the elimination of the current regulatory uncertainty, would establish a working market for slots. Most markets do not have auctions, but work and establish fair value without much difficulty. This secondary market would promote efficiency and competition by allowing carriers to sell and trade slots.

Further, we are advancing several major initiatives to enhance capacity at LaGuardia and the New York area airports including further airfield pavement projects to improve the efficiency of the airport, planning for a new central terminal building to increase gate size at

LaGuardia, and the funding the Stewart Airport rail study to enhance Stewart as a viable alternative airport. We remain committed to working with the FAA and key stakeholders in advancing the right approach for the New York region airports.

In: FAA Airspace Redesign…
Editor: Sidney R. Rothman

ISBN: 978-1-60692-710-6

Chapter 6

TESTIMONY OF EDWARD P. FABERMAN, BEFORE THE UNITED STATES HOUSE OF REPRESENTATIVES, COMMITTEE ON TRANSPORTATION AND INFRASTRUCTURE, SUBCOMMITTEE ON AVIATION REGARDING "CONGESTION MANAGEMENT IN THE NEW YORK AIRSPACE"

Edward P. Faberman*

Good morning Chairman Costello, Ranking Member Petri, and Members of the House Transportation and Infrastructure Subcommittee on Aviation.

I am very pleased to be here today to talk about issues that are critical to the nation's air carriers, communities throughout the country, and to the traveling public. My name is Ed Faberman and I am Executive Director of the Air Carrier Association of America. The ACAA and its low-fare carrier members are dedicated to providing affordable airfare travel options to all American travelers who receive significant benefits when low-fare options are available in their communities.

We thank this Committee for holding this hearing, for your dedication to enhancing the nation's ATC system, and for supporting the growth of air commerce and airline competition. We are at a point in time that if steps are not taken to ensure that the dream of deregulation will remain an essential part of U.S. airline service, we may see the disappearance of travel options for millions of travelers.

We note statements on these issues by committee members:

* Executive Director, Air Carrier Association of America

Chairman Oberstar

The House Subcommittee on Aviation (May 14, 2008) — "[A] reduction in competition [is] likely to lead to higher fares, a deterioration in service, and financially weakened survivors."

Hearing of Subcommittee on Aviation (March 18, 1999) "deregulation has saved the traveling public $12 billion a year. And the answer to much of the woes that we are hearing today is competition, not necessarily more regulation...The Department has to be vigilant to assure that there is competition for carriers, because competition is the best way to assure that they clean up their own act."

Ranking Member Mica

House Subcommittee on Aviation (July 12, 2006) — "It is clearly in the best interest of consumers for the invisible hand of the marketplace, not the heavy hands of Congress or federal bureaucrats, to set air fares and service options."

Subcommittee Chairmann Costello

"Competition is what is keeping prices reasonable today. If you eliminate competition, the fewer airlines you have, the fewer choices people have."

February 15, 2008 NBC News broadcast.

Subcommittee Ranking Member Petri

House Subcommittee on Aviation (April 25, 2007) "Overall, airline deregulation has brought better service and lower prices to the majority of communities around this Country."

These statements support the need for the Department of Transportation to take immediate steps to ensure that true competition will continue.

We appreciate the situation faced by Secretary Peters and Acting Administrator Sturgell in addressing delay and congestion issues that are once again plaguing the industry and disrupting the travel plans. We are glad to be discussing these issues alongside D.J. Gribbin and representatives of the Port Authority and other carriers. Unfortunately, these are not new issues.

What year was this statement made:

> "Throughout the year, there were strong indications that the airport and airway system was on the verge of saturation, especially within the so-called golden Triangle — Chicago, New York and Washington. Flight delays were far beyond mere annoyance and inconvenience. Time spent in holding patterns became all too frequent. For example, traveling between New York and Washington sometimes took as long as four hours. The airports serving all three of these cities and the airway system connecting them were simply congested during too many hours each day. This deplorable situation was uppermost in the minds of local and federal transportation officials, including DOT/FAA, and local airport operators. And in the minds of airline and their passengers. Everyone agreed that airport and airway capacity in critical areas of the nation needed to be increased, but there was no

consensus as to how soon this could be accomplished, or exactly what to do in the meantime."

THIS STATEMENT DESCRIBED CONDITIONS IN 1968!!

During the fall of 1968, the Secretary and the Acting Administrator of the FAA expressed the belief that governmental action limiting operations at some of the busiest airports would probably become necessary. Unfortunately, those restrictions have now been in place for 40 years.[1]

While we agree that it is important to reduce delays and congestion throughout the system and create a method of distributing airport capacity for the long term, immediate steps must also be taken to ensure that airline competition and deregulation do not become only memories of a system that used to exist. If immediate steps are not taken to strengthen competition, we will forever lose those critical options.

The airline industry is at a defining point in its history. We have recently seen several low cost carriers file for bankruptcy and cease all operations. At the same time, airline costs are out of control and they continue to rise. As this happens, barriers to entry increase.

While we are supposed to have open markets and a deregulated system, as a result of the significant increase in operations added by the dominant carriers at EWR, JFK, and LGA, all New York airports are once again closed to competition (ORD is also closed to entry). After closing these airports, the Department has not taken the necessary, corresponding steps to promote entry and competition.

Airports that are slot controlled have less low fare service and competition than most other airports in this country. Legacy carriers control 95% of the slots at LGA, EWR and ORD. At LGA, small carriers are limited to 10 roundtrips.

The one "closed" airport that has competition is JFK *because* the Department took specific action to provide JetBlue (formerly New Air) with significant entry at the airport when it first started its operation.[2]

When addressing closed airports, the Department has on multiple occasions emphasized the importance of preserving competition.

At the NY Scheduling Meeting (October 23, 2007), Secretary Peters stated:

> "I am not in favor of a system that limits competition, nor do I want to reduce the ability of new entrants to fly into New York."

FAA and DOT have repeatedly emphasized the importance of increased competition and frequently stated that establishing a competitive environment is a priority.[3] The LaGuardia Airport Congestion Management NPRM highlighted that there is *clear statutory authority* to promote competition when allocating slots/operating authorizations:

> Keeping available a variety of adequate, economic, efficient, and low-priced air services; placing maximum reliance on competitive market forces and on actual and potential competition; avoiding airline industry conditions that would tend to allow at least one air carrier unreasonably to increase prices, reduce services, or exclude competition in air transportation; encouraging, developing, and

> maintaining an air transportation system relying on actual and potential competition; encouraging entry into air transportation markets by new and existing air carriers and the continued strengthening of small air carriers to ensure a more effective and competitive airline industry. (Docket FAA-2006-25709, 71 Fed.Reg. 51360, Aug. 29, 2006.)

The "Notice of Alternative Policy Options for Managing Capacity at LaGuardia Airport and Proposed Extension of the Lottery Allocation," June 12, 2001 (Docket 2001-9852), stated:

> It is DOT's intention to develop a full array of public policy tools to develop a comprehensive aviation strategy that focuses on ways to reduce delays, improve airport capacity management, enhance competition, and promote the efficiency of the overall aviation system.
>
> The option should be flexible enough to allow policy makers the opportunity to address certain policy goals such as ensuring air carrier competition and service to small communities.

We are *not* operating in an environment where costs are stable — rather costs, including fuel, security and facility expenses, continue to significantly increase. Smaller carriers are not in a position to take extra steps at all airports to manage costs because they may have limited operations and facilities at those airports. Therefore, the costs for smaller carriers are higher at many airports. To manage those costs, it is essential that these carriers are allowed to expand their presence at these critical airports.

Although the Department has repeatedly stated that one of its primary objectives is to promote competition, it has not taken steps to increase the small percentage of flights operated by low-fare carriers at closed airports.

The current allocation system which is based on grandfathering of the slots while allowing each slot holding carrier to determine to whom and under what conditions they would lease or sell slots is anti-competitive.

As part of its "solution" to address congestion and entry at the New York airports, the Department has proposed a market approach that includes an auction. The comment period for the auction proposal closed this week, and various parties oppose the auction notice. Many questions remain about the Department's auction proposal. We understand the concerns raised by a number of parties with auctions.

We believe that it may be appropriate to consider market approaches as a longer term solution. As we review market approaches, however, immediate steps must be taken to make sure competitive options survive the current situation faced by the industry. As large carriers announce a reduction in operations at major airports, the elimination of aircraft from their fleets and possible mergers, now is the time to withdraw slots and facilities and make them available to carriers with small numbers of slots. Before a market mechanism is implemented, the Department must first make slots available to smaller carriers. To do this, the Department should withdraw some slots held by carriers holding more than seventy-five (75) slots as of January 1, 2008 and distribute those slots to limited incumbents operating aircraft with at least 110 seats that do not have codeshare arrangements or are partially owned by another carrier at the same airport that holds more than twenty slots.

We agree that market approaches should be considered if they promote competition. While the current Department Market Mechanism Notice would take slots from incumbents to redistribute them, it will not necessarily lead to the growth of competition. Allowing carriers with hundreds of slots to bid on slots against a carrier with 20 slots is unlikely to result in significant changes in slot ownership and will not increase competition. Moreover, it would be difficult for a carrier with a few operations at an airport to spend millions of dollars to outbid dominant carriers and still offer low fares.

After providing slots to smaller carriers, we do no object to the consideration of market mechanisms. To be considered, market approaches need to first enhance competition and any procedure to sell or lease slots must ensure that all carriers have a chance to bid on operating authorizations. Therefore, a blind mechanism must be implemented. As the Department of Justice has stated, incumbents with market power "will always have an incentive to outbid an equally efficient entrant for any slots offered." The Department commented:

> "Indeed, an incumbent with market power may well be able to outbid a <u>more</u> efficient entrant, simply because maintaining market power is more profitable than entering a competitive market...once a potential buyer's identity is known to the seller, the seller has every incentive to seek out an incumbent airline that would be willing to offer more money to maintain its market power than the entrant would be willing to apply to erode it."[4]

> The Department of Justice also noted:

> First, the mechanism chosen should ensure efficient allocation of scarce capacity at LGA, with slots going to their most productive use. Second, the mechanism chosen should permit competition to flourish. Allowing entrants to compete in a competitive primary market for the purchase of slots will increase the chances that more efficient entrants are able to acquire the slots they need. In order to promote a vibrant and competitive airline system, the scarce LGA capacity should be allocated in a way that encourages rather than discourages competition. Comments of the U.S. Department of Justice, June 20, 2002, Docket 2001-9854.

To *promote competition*, the following requirements must be implemented as part of any market approach:

- *First, provide slots to limited incumbents.*
- *Second, reserve some number of slots* only *for new entrants and limited incumbents.*
- *Third, establish a blind sale/transfer allocation system for any proposed sale or lease of slots.*

CONCLUSION

This Committee has played an active role in improving the nation's aviation system and in opening doors to competition and travel options for consumers. The ACAA applauds statements supporting competition made by Committee Members and on actions taken by the Committee. We thank you for holding these hearings. We are anxious to work with the

Committee and the Administration to address issues that significantly impact carriers, passengers and communities.

It is essential that we create a first rate system that makes flying easier and safer. At the same time we must continue the dream of deregulation. All actions taken must ensure that competition is not blocked. Some of the actions being taken to address delays and congestion could forever close the door on competition and prevent the growth of small carriers at major airports. Special care must be taken to ensure that this does not happen. Our dream of deregulation is to create a high tech, safe, "delay free" and secure system that maximizes consumer choices and ensures that low fares are available to all. Let's not make that dream just a memory. Instead, together let's make this dream become a reality.

We look forward to working with you on this and all matters. Thank you.

END NOTES

[1] FAA has on multiple occasions in the past addressed growing congestion problems. Unfortunately, the fixes developed have closed markets and limited competition. In 1968, the Acting Administrator of the FAA (Notice of Proposed Rule Making and Notice of Public Hearing; 33 FR 12580, September 5, 1968) stated:

> Delays of varying magnitude are encountered at many terminal areas... Congestion at these terminals frequently requires the imposition of traffic flow restrictions creating backup delays throughout the air transportation system. A reduction in air traffic delays can be accomplished only by increasing the capacity of the system or decreasing the demands placed upon it. Certain changes in air traffic and airport procedures and practices are already planned by the FAA to increase aircraft handling capacity.

FAA has on multiple occasions in the past addressed growing congestion problems. Unfortunately, the fixes developed have closed markets and limited competition. In 1968, the Acting Administrator of the FAA (Notice of Proposed Rule Making and Notice of Public Hearing; 33 FR 12580, September 5, 1968) stated:

> Delays of varying magnitude are encountered at many terminal areas...Congestion at these terminals frequently requires the imposition of traffic flow restrictions creating backup delays throughout the air transportation system.
>
> A reduction in air traffic delays can be accomplished only by increasing the capacity of the system or decreasing the demands placed upon it. Certain changes in air traffic and airport procedures and practices are already planned by the FAA to increase aircraft handling capacity.

On November 9, 1969, as a response to the delays in terminal areas, the FAA established the high density rule. When the rule was promulgated, the FAA stated, "the rule should be considered to be only a temporary solution." Unfortunately, thirty-nine years later the high density rule remains in place at Ronald Reagan Washington National Airport ("National") and La Guardia Airport ("LaGuardia") and may have to be reinstated at other airports to prevent continued delays.
The FAA addressed delays/congestion in 1984 and 1987:

> There can be little doubt that our nation's air transportation system is currently facing a substantial delay problem. According to DOT, airlines and travelers suffered 39,000 delays of more than 15 minutes in July of this year alone… It is premised on the view that a significant part of the delays stems from the hubbing practices of various carriers that results in the concentration of many flights within a short period of time every day... We tentatively concluded, however, that airline scheduling practices were a significant source of delays, since airlines appeared to be scheduling more flights at peak periods at major airports that could be accommodated by the available taxiways, runways, and airspace...Our show-cause order relied upon the excess schedules operated at Atlanta and Boston to show that airline scheduling practices cause much of the delay problem. [Docket 42410, "Application For Discussion Authority and Prior Board Approval of Carrier Agreements to Integrate Schedules," Order 84-8-129], "Order Granting Discussion Authority," August 31, 1984]

We find that the discussions are necessary to help alleviate the serious inconveniences caused the traveling public by the worsening delay problem, and that the discussions should therefore be approved and granted antitrust immunity. We are reluctant to authorize carrier schedule discussions, but the worsening delay problem demands that action be taken to help alleviate the problem. We believe that scheduling discussions are the only available altemative that can be {Docket 44634, "Discussion Authority For Carrier Agreements to Shift Schedules," Order 87-3-39, "Final Order Granting Discussion Authority," March 11, 1987]

[2] JetBlue has a major presence at JFK because in Order 99-9-11 (September 16, 1999), the Department granted New Air Corporation (now JetBlue) 75 slot exemptions at JFK. In that order, the Department clearly emphasized the importance of low-fare competition at airports dominated by legacy carriers:

The GAO's 1996 study, *Airline Deregulation: Barriers to Entry Continue in Several Key Domestic Markets* (the GAO Report), stated that "control of slots by a few airlines greatly deters entry at key airports in Chicago, New York and Washington," We made clear our support for increased competition and our willingness to invoke available tools to promote competition when we stated in our January 6, 1997, response to the GAO Report that "the Department intends to be more receptive to considering competition as a factor in granting slot exemptions to new entrants under the exceptional circumstances criterion."

As we noted earlier, many authorities, including members of Congress, have concluded that the High Density Rule is a serious barrier to entry, which has had a dampening effect on domestic airline competition.

JetBlue's business plan is to bring to the New York metropolitan area and many of its short- and medium-haul communities of interest a route system of price-competitive transportation services comparable to those that Southwest Airlines has brought to other cities throughout the nation. It is indisputable that Southwest has had a singularly positive effect on fare competition in literally every market it has chosen to serve. In many other markets, other low-fare new entrants have also had a salutary impact on domestic fares. Wherever those carriers have gone, fare competition has followed and traffic has increased, in many cases dramatically.

[3] Promoting entry and competition is also an essential mandate placed on the Department by Congress when it enacted the Airline Deregulation Act of 1978 49 U.S.C. 1301 — which charges the Department with facilitating new entry and competition in the airline industry. Under 49 U.S.C. § 40101:

(a)...the Secretary of Transportation shall consider the following matters, among others, as being in the public interest and consistent with public convenience and necessity.

(7) The prevention of unfair, deceptive, predatory, or anti-competitive practices in air transportation and the avoidance of

A. unreasonable industry concentration, excessive market domination, and monopoly power; and other conditions;
B. that would tend to allow one or more air carriers unreasonably to increase prices, reduce services, or exclude competition in air transportation.

(9) The encouragement, development, and maintenance of an air transportation system relying on actual and potential competition to provide efficiency, innovation, and low prices, and to determine the variety quality, and price of air transportation services.

[4] Comments of the United States Department of Justice regarding the Notice of Alternative Policy Options for Managing Capacity at LaGuardia Airport and Proposed Extension of Lottery Allocation, Docket No. FAA-2001-9854, p.6.

In: FAA Airspace Redesign and Congestion Management ISBN: 978-1-60692-710-6
Editor: Sidney R. Rothman

Chapter 7

TESTIMONY OF CONGRESSMAN CHRISTOPHER SHAYS "CONGESTION MANAGEMENT IN THE NEW YORK AIRSPACE" SUBCOMMITTEE ON AVIATION HOUSE COMMITTEE ON TRANSPORTATION AND INFRASTRUCTURE

Christopher Shays

Chairman Costello, Ranking Member Petri and the Aviation Subcommittee, I appreciate the Subcommittee's willingness to discuss congestion mitigation strategies for the northeast and look forward to a meaningful discussion about the benefits, drawbacks and unintended consequences redesigning the airspace will have on air travel, congestion, the environment and quality of life on the ground.

First, let me say I understand the Federal Aviation Administration's (FAA) desire to improve efficiency at LaGuardia, Kennedy, Teterboro, Newark and Philadelphia. Many business travelers are frustrated by the long delays at these airports.

Airline passenger volume continues to increase and the percentage of flights delayed 15 minutes or more in 2007 is close to surpassing the record set in 2000. Statistically, there was a significant increase in congestion and delay throughout the national aviation system in 2007. The situation has been especially noticeable at certain key airports, like JFK. Seventy-five percent of chronically delayed flights nationwide arrive and depart from New York airports.

The Bureau of Transportation Statistics reported between 1995 and 2008, the number of airline operations increased from 1,785,000 flights to 2,389,000 flights, or 25 percent. During that same time period, the percentage of on-time amvals dropped from 78 percent to 72 percent, while the percentage of late arrivals increased from 15 percent to 21 percent.

With that being said, I am concerned the FAA and DOT are apparently implementing two strategies for congestion mitigation irrespective of one another. The first is the proposed air space redesign, and the second is the imposition of flight caps at the nation's busiest airports. Furthermore, the FAA's continued refusal toconsider the impact market-based changes will have on the need to redesign the airspace is troubling.

Under the Integrated Airspace Alternative, the FAA's prefened airspace redesign plan, significant airplane traffic would be routed over parts of Fairfield County at the expense of the region's quality of life. Redding, Ridgefield, Weston, Wilton, New Canaan, Darien and Stamford would experience significant air traffic overhead because the redesign would shift the approach for flights to LaGuardia from the north.

The FAA has failed to provide any noise mitigation to our region, despite the wide swath of land over the Fourth Congressional District that will be adversely affected by planes flying at altitudes as low as 4,000 feet in the southern part of the district.

The FAA is also continuing to implement its redesign despite DOT'S auction proposal that could alleviate the need to redesign the airspace altogether.

Inadequate Noise Mitigation Strategies

The FAA is not required to present noise mitigation strategies even though there is significant impact on the region. Because there is no mandate for the FAA to consider quality of life, redirecting air traffic over previously unaffected areas is given the same weight in the agency's decision-making process as keeping traffic over areas that already have air traffic.

Even more concerning, no attempt has been made to utilize unpopulated, or less populated, tracts of land, industrial or commercial zones, major highway systems or large bodies of waters for mitigating noise impact, or to set minimum altitudes.

If the FAA had to consider the quality of life impacts of the Integrated Airspace Alternative, it would have never concluded this airspace redesign was the appropriate first attempt at relieving air traffic congestion.

It is unreasonable to design airspace without regard to impacts on the ground. For this reason, the FAA should adopt a "no action" alternative at this time until other congestion mitigation strategies, such as the auction proposal or moving peak flights to off-peak hours, have been thoroughly examined and disqualified.

Government Accountability Office (GAO) Study

On September 20,2007, Congressmen Rodney Frelinghuysen, Scott Garrett, and Elliot Engel and I offered an amendment to H.R. 2881, the FAA Reauthorization Act, requiring the GAO to assess the strengths and weaknesses of the FAA's environmental impact study, the costs and impacts of redesign, and whether the FAA followed due process in creating their proposal. It would be interesting to learn, for instance, if the GAO believes that what the FAA is proposing to do at LaGuardia and Newark would eliminate the need for a redesign.

I am hopeful this study, which is expected to be released on August 1 of this year, will reveal options as effective as redesigning the airspace to mitigate congestion that will not result in any new residents being affected by air-noise.

OTHER CONGESTION MITIGATION STRATEGIES

It seems to me there are other solutions that should be considered before implementing such a radical alternative that negatively affects so many thousands of residents throughout the Northeast.

The FAA has consistently stated the only way it can reduce congestion and travel delays is by redesigning the airspace. However, the case for the redesign is built on the flawed premise that delays are caused by the design of the airspace. This theory neglects to account for inadequate terminal, runway and gate capacity; insufficient Air Traffic Controller Staffing; weather; and over-scheduling flights. It also fails to account' for the impact market-based strategies could have on congestion.

What is particularly frustrating about the FAA's actions is the Department of Transportation's implementation of market-based solutions for congestion relief, while the FAA refuses to consider the impacts of these proposals on airspace congestion. We believe moving flights out of peak travel times and implementing slots and quotas at congested airports would have the same effect as a redesign, and must be considered.

I have to ask the FAA and this committee: is the airspace redesign the best we can do to mitigate airline congestion? In implementing the redesign, is it possible we are overlooking other market-based solutions to airline congestion? I believe the answers to those questions are no and yes, respectively.

The DOT'S imposition of flight caps at LaGuardia, Kennedy and Newark as a congestion mitigation strategy presents a unique opportunity to evaluate a congestion mitigation strategy that would not require redesigning the airspace. I believe redesign implementation should be postponed until we can determine whether these caps effectively reduce delays and congestion at three of our busiest airports.

I am interested to assess the effect the temporary cap has on passenger service and airline delays at Newark upon its expiration on October 29, 2008.

The FAA has also implemented an 83 flight-per-hour cap at Kennedy in New York and Newark in New Jersey. Last year, Kennedy handled about 100 flights per hour and Newark about 95 flights per hour at peak times. The caps will remain in place at Newark and Kennedy for two years, which will provide an opportunity to compare flight delays and cancellations.

To generate revenue for additional market-based congestion mitigation strategies, the DOT has proposed an auction system for flights during peak hours whose profits would be placed in the General Aviation Fund and used for mitigation activities like taxiway expansion and Air Traffic Control System (ATCS) upgrades. The auction system will provide incentives for airlines to operate more flights during off peak hours.

For example, if a peak travel time occurs between 7 a.m. and 9 a.m. at LaGuardia, it seems to me we should encourage airlines to schedule flights between 5 a.m. and 10 a.m. to reduce the volume of passengers traveling at once.

Simultaneously, we need to examine the schedules of airlines operating out of overbooked metropolitan airports and adjust their arrival times to account for predictable delays.

For instance, on a Monday morning at LaGuardia, there are regularly more than a dozen commercial flights on the taxiway waiting for clearance to take off. Since the scheduled departure times for these flights are known in advance, I believe it is reasonable require the airlines to provide a more accurate estimated time of arrival.

Finally, I am concerned that increasing capacity in the air will not matter if we don't provide additional terminals and runways to get more planes off the ground. The bottom line is there is a finite number of planes that can safety take off and land in a given period of time.

Growing Opposition

In addition to an inadequate assessment of alternative congestion mitigation strategies, it seems to me the FAA has overlooked the growing opposition of residents on the ground.

A petition circulated throughout Connecticut's Fourth Congressional District has garnered the support of 5,500 residents. The petition stated, "We believe that the true social costs of the FAA's proposal have not been adequately evaluated, particularly the increased noise levels, which will cause irreparable hm to our communities."

Jim Gerlach (PA), Rob Andrews (NJ), Joe Sestak (PA), Scott Garret (NY), John Hall (NY), Eliot Engel (NY), Rodney Frelinghuysen (NJ), Nita Lowey (NY) and I wrote FAA Acting Administrator Robert Sturgell on March 12 urging the FAA to postpone the implementation of the airspace redesign. Concern is mounting from the tri-state area.

Our letter stated, "If the GAO finds significant flaws in this plan or the methodology used to develop it, Congress should have an opportunity to work with the FAA to address these concerns before its implementation its irreversible."

The bottom line for me is the FAA is using a flawed process to reach an even more flawed conclusion. We cannot simply ignore the impacts of air noise on the ground. Subsequent attempts to measure and then mitigate the hugely negative impacts are inevitably inadequate. We should all take a step back here and look at the big picture, and then focus on the less radical alternatives before subjecting millions more to the burden or air noise.

Thank you for your time and consideration of the unintended consequences the redesign of the New York/ New Jersey/Philadelphia Northeast Airspace will have on the ground.

In: FAA Airspace Redesign and Congestion Management
Editor: Sidney R. Rothman
ISBN: 978-1-60692-710-6

Chapter 8

United States Senator Charles E. Schumer, Remarks as Prepared, House Subcommittee on House Subcommittee on Aviation: Congestion Management in the New York Airspace

Charles E. Schumer

First, I'd like to thank Chairman Costello and the Aviation Subcommittee for holding this hearing today on such an important issue and allowing me to speak.

Anyone who has traveled in the last year has felt the pain and frustration that now defines air travel in the U.S. Delays and cancellations have crippled the entire air traffic system leaving passengers stranded across the country. Time and time again, we've seen the Department of Transportation attempt to address New York's congestion problems using ideological, market based theory, and other untested experiments.

First it was caps, then it was congestion pricing, and now all we're hearing about are auctions. Instead of focusing on real, tested solutions to solve this problem, like upgrading decades old technology and hiring an adequate number of controllers to staff New York's towers, the DOT continues to miss the point. To make matters worse, the FAA's treatment of controllers has lead to an unprecedented rash of retirements thus compounding the problem.

The bottom line is that at the current pace and under the current plan, within the next five years the New York City airspace will be in total gridlock. The wave of controller retirements and the FAA's inability and unwillingness to upgrade antiquated technology will simply cause a meltdown in New York and across the country.

Last summer's travel season was hampered by some of the worst delays ever recorded by the FAA — nearly 30% of all flights were delayed or cancelled. We all remember countless reports from travelers who were stranded on the tarmac for hours on end, and whose flights were cancelled, forcing them to miss out on long-awaited summer vacations or important business meetings. In a report issued last month by the Joint Economic Committee, of which I am the Chair, we found that these flight delays cost passengers, airlines, and the U.S. economy over 40 billion dollars. Sadly, this is just the beginning

To make matters worse, air traffic control towers across the country are dangerously understaffed. When you look at the workforce nationally, most facilities in the country have staff levels that are more than 35% trainees. In December of 2007, developmental trainees comprised 25% of the national controller workforce, up from about 15% in 2004. Given that the trainees cannot staff a controller position without a fully certified controller supervising, these numbers are alarming.

In New York, when you count fully certified controllers only, excluding trainees, the JFK tower is only 60% staffed, having 22 full time controllers when it should have 37. The LaGuardia tower is only 60% staffed, having 22 full time controllers when it should have 36. The Newark tower is only 67% staffed, having 27 full time controllers when it should have 40. The FAA will give you higher staffing numbers, but that's because their numbers are doctored to include trainees who cannot staff a position without supervision.

At the New York Center, which manages traffic entering into the New York airspace, the average training time at New York Center is slightly more than four years, the second highest in the country. However most of the 71 controllers in training have not even started their training. Since September of 2006, this facility has lost 57 fully certified controllers and only seen 18 hired to replace them. Similarly, the New York TRACON, responsible for all arriving and departing traffic, has lost 23 fully certified controllers to retirement and resignation, and only gained one just this month.

Nationally, the culture of fear created by the FAA has led 3,300 controllers to leave the workforce since 2005 with 15,500 more expected between 2008 and 2017. In just six months from October of 2007 through March of 2008 nearly 1,000 controllers left the workforce — half of them to retire. And of the FAA's 1,800 new hires in FY07, only 150 of them have been fully certified. As the recent DOT IG report pointed out, the FAA may be hiring new trainees, but they aren't adequately training them to become fully certified professional controllers. It doesn't help to include trainees in your staff counts when these controllers can't work unless a full time controller is overseeing them.

This wave of retirements we are seeing just sets up a vicious cycle where there will be more overtime needed, then quicker and more retirements as workers get burnt out. In other words, the next wave of retirements will make this one look like nothing The bottom line is that New York's towers are understaffed, controllers are retiring in droves, and our air traffic controllers continue to operate with equipment that over 50 years old.

If the DOT and FAA don't take steps upgrade the technology, improve capacity at New York's airports and hire, train and retain more full time controllers, then all we can look forward to is even more congestion and problems.The key is enhancing capacity at the airport and fully staffing the control towers.

Now, the DOT plans to auction off takeoff and landing time slots at New York City's three area airports. This will in NO WAY reduce congestion in New York's airspace. Instead, this plan could limit consumer choices and have a dire impact on service to small communities. And I will act quickly to prevent this plan from being implemented.

As travelers, we feel the burden of an airline industry gasping to catch its breath, and the worst possible thing the Administration could do to passengers and the industry in general is to deliver one big punch in the gut by auctioning off slots at New York's airports.

As I've said multiple times before, this is an ideological, untested experiment from those in an Ivory Tower. There is no proof that DOT's auctions plan will do anything to reduce congestion, yet DOT insists that this market-based solution is the only effective proposal. The DOT's misguided plan to sell take-off and landing slots to the highest bidder won't make your plane take off faster, it will just cost New Yorkers more to fly and throw LaGuardia, JFK and Newark into chaos.

Auctions have no proven track record of working. They have never been tried in any U.S. airport, nor have they been tried at any airport in the world. In fact, the entire plan is modeled after London's road congestion pricing plan. I don't know of one expert who believes this plan will work or is beneficial to the consumer. In fact, the Air Transport Association which represents the airlines, the Port Authority of NY and NJ and the Airports Council International, which represents the airports across the country, are ALL opposed to slot auctions. In fact, their plan is unconstitutional and will be challenged in court.

Airports in the New York City area, which are some of the world's busiest, should not be the guinea pigs in DOT's harebrained congestion experiment. When I spoke to Acting Administrator Sturgell, I told him that I would oppose any experiment in the New York area involving auctions, and because he couldn't give me a straight-forward answer as to the FAA's plans, I placed a hold on his nomination which still exists today.

Auctions will also be detrimental to consumer choices. Small communities in New York State and across the country, like Buffalo and Syracuse, will inevitably see a decrease in service when airlines abandon smaller routes in favor of more lucrative long-distance and overseas flights. Under this proposal consumers will most likely see low-cost air service cramped out of the market and additional expenses airlines incur to buy slots will inevitably be passed along to consumers. This is the last thing consumers need.

This auction plan is further flawed by the assumption that all slots are interchangeable. However, we all know that a flight from JFK to London is not interchangeable with a flight from JFK to Buffalo. Long-distance and international flights need to take off and land at certain times of the day, and large planes need specifically-designated large gates and terminals.

The DOT claims that auctions are part of a market-based solution to address congestion, but as you can see auctions will do nothing more than hurt consumers by and make our domestic airlines less competitive.

So, today I have introduced legislation supported by both the airlines and the airports including the Port Authority of New York and New Jersey, which will protect travelers by prohibiting the DOT from implementing auctions in New York City or at any commercial airport in the country. I cannot stress enough how important this legislation is. The DOT insists that auctions are the only viable solution to congestion problems in New York but as you can see, this is no solution at all. My bill protects the proprietary rights of the airports while preserving the airlines ability to offer consumers the best variety of air service they can provide.

Once again, the DOT is putting ideology before efficiency and the almighty dollar before New Yorkers and passengers around the country. It is time to put an end this scheme once and for all.

In: FAA Airspace Redesign and Congestion Management
Editor: Sidney R. Rothman
ISBN: 978-1-60692-710-6

Chapter 9

STATEMENT OF THE HONORABLE D.J. GRIBBIN, BEFORE THE UNITED STATES HOUSE OF REPRESENTATIVES, COMMITTEE ON TRANSPORTATION AND INFRASTRUCTURE SUBCOMMITTEE ON AVIATION CONCERNING AVIATION CONGESTION MANAGEMENT

*D.J. Gribbin**

INTRODUCTION

Mr. Chairman and members of the Subcommittee, thank you for the opportunity to testify again on the Department of Transportation's (DOT) continuing efforts to address aviation congestion.

Status of the Industry

Before I go into detail about the DOT's efforts to address congestion, I want to take a moment to talk about the particularly challenging environment currently facing the airlines. As my fellow panelists well know, record oil prices, a slowing economy, and increased competition are just a few factors that have created a number of significant challenges for airlines – challenges that certainly will change the face of the aviation industry in the years to come.

To meet these challenges, many carriers are raising fares, streamlining operations, and reducing service. It is possible that some of these measures will result in reduced congestion – however, so far we have yet to see widespread evidence of carriers pulling out of the busiest

* General Counsel, U.S. Department of Transportation

(and most congested) airports. Although, Continental announced just last week that they are eliminating service to 15 communities, it is likely that the busiest and most congested airports will not see an overall reduction in service – and even if there is a reduction, history tells us that the aviation industry is very cyclical and that service will return to – and exceed – the record levels we saw last year.

In 2007, the aviation industry recorded the second worst year for delays since 1995; 27% of flights were delayed or cancelled in 2007. Both the frequency and the severity of ground delays were unprecedented. The costs of delays are huge – the Senate Joint Economic Committee estimates that last year flight delays alone cost passengers, airlines, and the U.S. economy over $40 billion. Additionally, the Travel Industry Association estimates that air travelers avoided over 41 million trips last year – leading to lost revenues and taxes of over $26 billion.

The cost of delays and congestion to the U.S. economy is huge and that is why, even if carriers reduce flights this summer enough to reduce congestion, we still must do something to fix the problems that caused last summer's horrible delays. We simply cannot wait until there is another summer of record delays before we do something to fix the system. That is why the Department will continue working on its initiatives to address congestion and introduce competition at capped airports.

LaGuardia/JFK/Newark Background

As you all know, the Department recently published notices of proposed rules intended to manage congestion at LaGuardia Airport (LaGuardia), John F. Kennedy International Airport (JFK), and Newark Liberty International Airport (Newark). We believe these proposals will ultimately provide travelers with more reliable service while maintaining competition among the many carriers in a vibrant New York market.

Congestion at these three New York airports is not a new phenomenon. Since 1969, the High Density Rule (HDR) has effectively capped LaGuardia to a limited number of operations per hour and capped JFK during its peak hours. Although Newark was once subject to the HDR, the FAA suspended its application in 1970 due to the fact that capacity was meeting demand. In recent years, however, operations have bogged down to the point where Newark is now one of the most delay-prone airports in the country. Current and anticipated demand during peak hours at all three airports approaches or exceeds runway capacity, causing volume-related delays, which can be aggravated by weather or other operating conditions. Operational improvements have not increased the capacity of the New York area to a point where the unconstrained demand for air service can be met without excessive congestion. Therefore, for now, all three airports are capped.

Straight caps without some mechanism to ensure an efficient allocation of scarce slot resources is economically inefficient and, therefore, not our preferred option. Our preference is to see airports address their challenges locally through implementation of capacity enhancing projects or procedures, whenever possible. However, the federal government will be involved once a congested airport impacts the rest of the national airspace. In this case, New York air congestion causes delays throughout the U.S., so the federal government cannot ignore the problem. Given the urgent need for action, caps were necessary at the New York City area airports.

When we consider economic regulatory issues, the Department has a statutory obligation to place maximum reliance on competitive market forces and on actual and potential competition. We know, however, that caps hinder the ability of air carriers to initiate or expand service at capacity constrained airports. Therefore, when seeking a solution to the aviation congestion issues that we currently face in the New York area, the Department must act to both promote competition by permitting access to new entrants, and to recognize the long-term investments in airports made by existing carriers. We do not believe that a simple imposition of caps without some mechanism to preserve competitive market forces benefits aviation consumers or the airlines.

With this in mind, we have set forth proposals for the New York area airports that we believe would reduce congestion the smartest way—by using market incentives to assist in the efficient allocation of airspace. Although market-based mechanisms are the most effective way to allocate scarce resources—like slots—we have taken a very conservative approach to introducing these mechanisms with this proposal. The vast majority of hourly operations at the airport, as much as 90 percent or more, would be "grandfathered" and leased to the existing operators for non-monetary consideration. The market-based aspect of our proposal involves auctioning off leases for only a limited number of the remaining slots.

Are There Alternatives to Caps and Auctions?

Expanded Capacity

Some have incorrectly suggested that expanding capacity should be the only government response to congestion in New York City and around the country. This view largely ignores the tremendous short-term opportunities to utilize existing capacity more efficiently. It also ignores the physical, economic, and political constraints on capacity expansion in many parts of the U.S. aviation system.

The Department shares the view that expanded capacity is a critical component of the long-term solution to relieve congestion and get travelers to their destinations on time and in a humane fashion. We are intensely focused on such solutions, both at the FAA with implementation of the Next Generation Air Transportation System (NextGen) and at the Department level. The FAA is hard at work bringing new technology and techniques online to unsnarl air traffic delays, and we appreciate the funding Congress has appropriated for these purposes. In recognition of these critical enhancements, the President's FY 2009 Budget Request would more than triple the investment in NextGen technology – providing $688 million for key research and technology to help meet the nation's rapidly growing demand for air travel, including the transformation from radar-based to satellitebased air traffic systems.

The FAA will begin rolling out several elements of the NextGen system this summer. This rollout will include the national debut of Automatic Dependent Surveillance-Broadcast (ADS-B) technology in Florida.

The FAA has chosen Miami as the key site for the installation and testing of Traffic Information Services – Broadcast (TIS-B) and Flight Information Services – Broadcast (FIS-B). These broadcast services are the transmission of weather and traffic information to the cockpit of properly equipped aircraft. In order to provide the services in roughly the southern half of the state, the contractor, ITT will install and test eleven ground stations in this area, including five at airports (Lakeland Linder Regional, Dade-Collier, Florida Keys Marathon Airport, Boca Raton Airport, and Sebastian Municipal).

The ITT installed equipment is currently undergoing a Service Acceptance Test (SAT) which began in May. In November 2008, the agency expects to commission (the FAA calls this an In-Service Decision or ISD) these broadcast services (TIS-B and FIS-B). Following the successful completion of ISD, the FAA can exercise an option in the ITT contract to deploy the services nationwide

The transition to ADS-B technology will allow the nation's air traffic control system to change from one that relies on radar technology to a system that uses precise location data from a global satellite network. Over the next few years, the FAA will also install and test ADS-B for use in Air Traffic Control Separation Services. The key sites for this initiative are Louisville, Philadelphia, the Gulf of Mexico, and Juneau. The FAA plans to commission the ADS-B services in September 2010 and complete a nationwide rollout by 2013.

The FAA also recently completed stage 1 implementation of its Airspace Redesign Project for the New York, New Jersey, Philadelphia area. The goal of the Airspace Redesign Project is to enhance the efficiency and reliability of the airspace structure and the air-traffic control system for pilots, airlines and the traveling public. The project modernizes the structure of the air traffic environment in an environmentally responsible manner, while laying the foundation for NextGen. Moreover, it will help to accommodate growth while enhancing safety and reducing delays. While airspace redesign will provide greater efficiencies and some congestion relief, it is not a complete solution.

The Department looks to increase capacity both in the air and on the ground whenever possible. Our support for expansion of O'Hare International Airport is one concrete example. The fruits of these efforts became clear on Monday when the FAA announced that it would allow the flight caps put in place at O'Hare in 2004 to expire because of the additional capacity the airport will gain from its new runway. Capacity increases must be part of the solution, particularly since we expect demand for air travel to resume its robust growth over the coming decade, despite the current temporary pause due to economic conditions. This is especially true in the nation's busiest metropolitan areas. However, capacity increases, both physical and operational, often take a long time to implement and may be limited in scope. Sometimes physical capacity cannot be expanded, such as at LaGuardia Airport. Operational improvements can help to address congestion, but sometimes they cannot provide enough capacity to meet demand. For example, in New York, even with the implementation of all the operational improvements initially suggested by the Air Transport Association (ATA) and the Port Authority, congestion was expected to double this year, assuming the FAA took no further action and the airlines moved forward with planned increases in their schedules.

There are additional solutions. Basically, we have a choice between two fundamentally different approaches – administrative remedies and market-based solutions. We believe that outdated government policies relying on administrative remedies alone have led to an inefficient allocation of the airspace, and that moving towards a system that includes market-based solutions will reduce these inefficiencies and contribute to an improved flying experience for air travelers.

Administrative Allocation

Instituting administrative remedies, such as caps, is an effective, but not efficient way to reduce delays. Limiting the number of flights into an airport will reduce congestion at that airport. The Department decided to institute a short-term cap at JFK and Newark airports because something needed to be done to avoid a repeat of the flight delays that we experienced last summer. However, caps are not the best solution for improving travel options for passengers.

Airlines are often enthusiastic in their support of caps at an airport they already serve. When a cap is established, incumbents are protected because they typically maintain their market share and the potential for new competition is diminished. The incumbent airlines' support for such a policy makes sense, because limited competition makes them more profitable and protects them from new entrants that might want to compete by offering lower fares.

Although caps protect existing airline business, they also prevent airlines from adding capacity at an airport unless they are able to obtain a slot from a competitor. As a result, one of the best-known problems with slots is that they encourage airlines to "babysit" slots, i.e., underutilize the slot by flying multiple small aircraft into an airport to maximize the number of slots an airline can occupy at the lowest possible cost.[1] As a result, slots do not always go to those who value them the most and who will use the capacity in the most efficient manner.

This limitation on capacity and competition naturally leads to fare increases at an airport, because it creates a scarce commodity, and passengers pay a premium for that commodity.

If caps are not the long-term answer, then the question arises – what is the solution?

Market-Based Remedies

Alfred Kahn, an airline economist and former Chairman of the Civil Aeronautics Board, said: "Whenever competition is feasible, it is, for all its imperfections, superior to regulation as a means of serving the public interest." Secretary Peters echoed that sentiment when she said: "Our preference is to find a way to let market incentives do the job, and not to return to the days of government-regulated flights and limited competition." Although the Department instituted caps as a short-term measure, we continue to explore market-based remedies as a long-term solution to congestion.

It is clear that the current system does not allocate airspace capacity efficiently. Solving that problem, however, should not entail government picking "winners and losers," particularly when, as currently structured, everyone involved in air travel feels like they are the loser—both those getting terrible service and those getting blamed for providing terrible service.

Market-based pricing has been demonstrated time and again as the most effective way to allocate a scarce resource that is in high demand. Space in a movie theater, use of cell phone infrastructure, or flights during certain times to certain destinations are all examples that illustrate that such pricing works. Pricing can balance demand with available capacity, resulting in less congestion and more reliable schedules. Also, pricing sends better signals as to where the system needs extra capacity, and it can supply the revenues to add such needed capacity.

Changing from the traditional, increasingly inefficient administrative controls to a market-based system has generated a fair amount of concern, primarily from the airlines. Change is difficult, and the airlines' concerns are understandable. In fact, very similar arguments were made by the airlines in opposition to deregulation. Concerns were raised about disruption to the industry, lack of a track record, and disruption to business models. However, the ATA Airline Handbook includes a long list of benefits that resulted from deregulation. The Handbook notes that deregulation stimulated competition, led to rapid growth in air travel, and reduced fares by more than 50% in real terms. We believe that market-based remedies directed at congestion will improve airline service like deregulation did.

Why Caps Must Be Combined with Auctions – and How It Will Result in Lower Fares

Implementing caps without any additional market-based mechanism for encouraging competition only increases the cost to consumers, since a lack of competition keeps fares high. A March 2001 Government Accountability Office (GAO) report found that "dominated markets tend to have higher airfares than airports that have more competition from other airlines." Fares in dominated markets averaged 41 percent higher than in markets where there was aviation competition. The difference in fares is largely attributed to the exclusive access granted to incumbent airlines and the incumbent airlines' ability to prevent new entrants from gaining entry to create a competitive market. Instituting slots without a market-based mechanism creates just this exclusivity of access by granting extensive landing rights to incumbent airlines and barring new, competitive entrants into the country's busiest airports.

Granting slots without market-based mechanisms creates a system where incumbent airlines fight to maintain large shares of the airport traffic and to limit the ability of lowcost carriers to compete. The 1996 DOT report *Low Cost Airline Service Revolution* details this anticompetitive culture at capped or dominated airports. The report identifies slot hoarding as one of the key characteristics of such a culture. Federal regulations require airlines to use their slots at least 80% of the time in order to retain possession of them. However, by splitting up larger flights into smaller ones ("downgauging") or by setting up a rotating schedule, airlines have unnecessarily taken up more slots than they would require to competitively serve their customers. Slot hoarding prevents new entrants from taking available slots and increases airplane throughput without increasing passenger throughput, adding greatly to congestion. The report maintains that the high fares charged at these dominated airports create incentives for an airline to use anticompetitive measures to discourage new entrants.

Using the historical backdrop of slots as a guide, we believe that integration of a market-based system into the proposal for slot caps is necessary to protect consumers and a competitive market. Estimates from the DOT's 1996 report valued savings from new entry competition at 35 percent for round-trip flights and 40 percent for one-way flights. A case-specific study on the effect of Southwest Airlines noted that with the opening of just one route between Oakland International Airport and Ontario International Airport in Los Angeles, fares dropped 60% and traffic tripled, increasing both passenger throughput as well as savings for consumers. Even nearby airports not directly offered service experienced a decrease in fare costs of up to one-third. Southwest is just one example of low-cost carriers whose entry into the market drove down prices and increased passenger throughput at previously dominated airports.

What Have We Proposed?

Last month, the FAA published notice of a proposed rule that would replace the orders imposing operating limits at JFK and Newark and establish a new rule limiting operations at these airports. Instead of reliance on repeated piecemeal approaches to limit and manage operations at JFK and Newark, we believe a better course is to adopt a longer-term rule dealing with the congestion and delays that we expect to persist at those airports. Although we continue to work toward capacity improvements, this proposal will complement capacity enhancement efforts.

Like the proposal for LaGuardia that I discussed the last time I was before this Committee, this proposal recognizes that a simple imposition of caps without some mechanism to ensure preservation of competitive market forces is inadequate. While this proposal is similar to and intended to mesh with the LaGuardia proposal, neither is reliant on the other for final action.

Under the proposal for JFK and Newark, all airlines operating at Newark and JFK would be given up to 20 slots a day for the 10-year life of the rule. The proposal offers two options for JFK. Under the first, 10 percent of the airline's slots above the 20-slot baseline would be made available via an auction. The revenue from those auctions would then be invested in congestion and capacity improvements in the region.

Under the second option for JFK, the airlines would auction 20 percent of slots above the 20-slot baseline and keep all of the proceeds. Depending on the option, between 91 and 179 slots at JFK would be affected out of 1,245 total slots at the airport.

The proposal also calls for auctioning 10 percent of slots at Newark Airport above the baseline annually for the first five years of the rule. As a result, only 96 slots out of a total of 1,219 slots at the airport would be auctioned over the 10-year span of the proposal.

As with any pricing plan pursued by the Department, this proposal for JFK and Newark complies with our international obligations and will not competitively disadvantage domestic carriers. Under this proposal, foreign carriers and domestic carriers are treated the same.

As with the LaGuardia proposal, under this proposal, airlines operating at the two airports would receive a 10-year interest in some of the world's most valuable aviation assets, free of charge, free of question, and free of hassle. Additionally, this proposal – just like the LaGuardia proposal – increases competition by creating a robust secondary market for trading of slots and allowing a way for new entrants to gain entry into a restricted airport.

Conclusion

Mr. Chairman, I appreciate the opportunity to explain to you our proposals for the New York-area airports. We are firmly committed to the idea that any long-term solution to mitigate congestion in the Nation's airspace must include a market-based mechanism. Caps alone have proven to be insufficient, and perpetuating the kinds of delays we experienced in the summer of 2007 is not tolerable.

I would be pleased to provide you or your staff with any additional information that might help explain our proposals and I would be happy to answer any questions you might have.

END NOTES

[1] GAO report GAO/RCED-99-234 notes on p. 16 that "For example, because the regulations allow a slot to go unused for up to 20 percent of the time, a carrier with five slots in 1 hour must operate only four flights in that hour on any day to obtain 80-percent use for each of its five slots. The carrier is allowed to "rotate" its four flights across the five slots over the 2-month period to prevent FAA from withdrawing the slot. The practice of a carrier's rotating actual flights among its allocated slots is commonly referred to as 'babysitting.' FAA officials emphasized that babysitting is not prohibited by existing regulation, provided that a slot meets the minimum-use requirements." See http://www.gao.gov/archive/1999/rc99234.pdf

Chapter Sources

The following chapters have been previously published:

Chapter 1 – This is an edited, reformatted and augmented version of a United States Government Accountability Office publication, Report GAO-08-786, dated July 2008

Chapter 2 - This is an edited, reformatted and augmented version of a Congressional Research Service publication, Report RL34284 dated December 14, 2007.

Chapter 3 – These remarks were delivered as testimony given on June 18, 2008. James C. May is President and Chief Executive Officer Air Transport Association of America, Inc., Before the Subcommittee on Aviation of the Committee on Transportation and Infrastructure: Congestion Management in the New York Airspace.

Chapter 4 - These remarks were delivered as testimony given on May 6, 2008. Douglas E. Lavin is Regional Vice President (North America), International Air Transport Association to the Committee on Transportation and Infrastructure's Subcommittee on Aviation, United States House of Representatives.

Chapter 5 - These remarks were delivered as testimony given on June 18, 2008. William R. Decota, Aviation Director, The Port Authority of New York and New Jersey to the United States House of Representatives, Transportation and Infrastructure Committee, Subcommittee on Aviation.

Chapter 6 - These remarks were delivered as testimony given on June 18, 2008. Edward P. Faberman, Executive Director, Air Carrier Association of America Before the United States House of Representatives, Committee on Transportation and Infrastructure, Subcommittee on Aviation Regarding "Congestion Management in the New York Airspace."

Chapter 7 - These remarks were delivered as testimony given on June 18, 2007. Congressman Christopher Shays to the "Congestion Management in the New York Airspace" Subcommittee on Aviation, House Committee on Transportation and Infrastructure.

Chapter 8 - These remarks were delivered as testimony given on June 18, 2008. United States Senator Charles E. Schumer, to the Remarks as Prepared, House Subcommittee on Aviation: Congestion Management in the New York Airspace.

Chapter 9 - These remarks were delivered as testimony given on June 18, 2008. Honorable D.J. Gribbin General Counsel, United States Department of Transportation Before the United States House of Representatives Committee on Transportation and Infrastructure Subcommittee on Aviation Concerning Aviation Congestion Management.

INDEX

B

C

D

E

F

G

H

I

J

L

P

Q

R

S